The GHOST WALKER

Wildlife in Canada (1966)
The Place in the Forest (1967)
Where the Water Lilies Grow (1968)
The Poison Makers (1969)
Cry Wild (1970)
Maple Syrup (1971)
Wildlife in North America: Mammals (1974)
Wildlife in North America: Birds (1974)
Paddy (1977)
The North Runner (1979)
Secret Go the Wolves (1980)
The Zoo That Never Was (1981)
Voyage of the Stella (1982)

The GHOST WALKER

R·D·LAWRENCE

Holt, Rinehart and Winston   New York

Published by Holt, Rinehart and Winston,
383 Madison Avenue, New York, New York 10017.

Library of Congress Cataloging in Publication Data
Lawrence, R. D., 1921–
The ghost walker.
1. Pumas. 2. Outdoor life—British Columbia.
3. Mammals—British Columbia. I. Title.
QL737.C23L38   1983   599.74'428   82-12111
ISBN 0-03-061594-1
First Edition

Designer: Susan Mitchell

Printed in the United States of America
1 3 5 7 9 10 8 6 4 2

ISBN 0-03-061594-1

For Ron and Kay Poulton,
close friends who give comfort
when needed and always welcome
the wanderer on his return.

The GHOST WALKER

 SITTING BEHIND THE PILOT AS THE TWO-seater aircraft leveled beside a snow-clad peak, I couldn't help thinking about the money I would waste if we failed to find a mountain lion within the rugged wilderness we were scrutinizing. I was fresh from a six-month sea journey and was now intent on studying this fascinating North American cat in British Columbia's mountain country. But first I had to locate at least one lion in the vast hiding place that sprawled below. *What were my chances?*

The region I had rather arbitrarily chosen to search lies within that part of the Selkirk Range that is bounded by the northern limits of the Columbia River. This is a country of tall peaks, icefields, bare granite, and dense forests, latticed by virtually uncountable creeks that weep their icy tears down the flanks of every mountain.

Occasionally, during the start of the flight, when the pilot had to climb high to clear a majestic peak, I was afforded a view of the Columbia, which, flowing northward from its Canadian source, turns abruptly at Mica Creek, then continues its journey to the Pacific Ocean, emptying itself, after having run for 1,215 miles, in the estuary that divides Oregon and Washington.

As I continued to worry, the aircraft threaded between passes, flew into a valley, and then rose to skim over white mountain tops that stretched from horizon to horizon, or were cloaked with mist in the far distance. The nearest ones, no more than two hundred feet away, presented a barren and desolate mien, yet were made beautiful by a pristine layer of snow or by patches of ancient blue ice that sparkled like gigantic jewels.

Forcing myself to turn my gaze from those impressive and rather intimidating peaks, I concentrated on the ground, searching it carefully with field glasses when we were flying high and with naked eyes when we descended to the lowlands. Occasionally I thrust my head out of the port or starboard windows. Since the passenger seat of this small plane was behind the pilot, I could look out both windows without having to lean across the flier. When one of the windows was open, the icy wind struck my face and made my eyes water slightly; but this was nothing compared with the blast I received when I put my head out!

When we had been searching for more than three hours without seeing a puma, I began to feel despondent. At one hundred dollars an hour, I had decided that I could afford only four hours of flying, and now the time limit was only fifty minutes away. I had seen a number of deer, a small group of woodland caribou, and two mountain goats during the first hour. Later, in one valley, I sighted six timber wolves and a large grizzly. I also saw five black bears, each in a different location.

Although the pilot was sympathetic by now, his manner indicated that he thought only a lunatic would be willing to spend so much money on the off-chance of seeing a mountain lion. Then my luck changed. The aircraft had just started to bank to port on my instructions, so I could get a better look at the terrain, when, with startling suddenness, a lithe, tawny shape emerged from a clump of young evergreens. A mountain lion! The cat raced straight across a scree slope, then stopped almost as quickly as it had appeared. Evidently the large golden animal had been startled and probably somewhat disoriented by the droning of the aircraft's single engine.

Excited, I leaned out the Aeronca's open window and the buffeting wind almost drove one of the eyepieces of my field glasses into my eye. This made me lose sight of the cat for a moment, and when I saw it again, it had turned back toward the shelter of the trees, but now hesitated anew, undoubtedly realizing that the small evergreens offered scant cover.

Reversing itself for the second time, the lion set off at a fast run, stretching its body as its legs covered the ground, the extra-long tail, characteristic of the species, flying straight out behind, its black tip bobbing slightly as the animal dashed over the treacherous rock fragments and sent small and dusty avalanches rolling downhill. The pilot, obeying my shouted instructions, tipped the small twin-seater to the right, giving me a better view of the speeding cat.

From three hundred feet up, I had an almost perfect view of the animal. Its ears were pressed back against its round head and its muscles and sinews moved rhythmically, alternately bunching and relaxing while its low-hung belly brushed over some of the larger stones on the lower flank of the mountain. Twice the puma looked upward and sideways before it reached the shelter of the tall trees that hid it from my sight, but although I had been given but a few seconds in which to watch the big cat, I was more than satisfied and content as the pilot leveled the Aeronca and I pulled my head back into the cabin. I would naturally have enjoyed watching the lion for a longer time, but the sight of it, however brief, had concluded to my satisfaction the first stage of my expedition.

Ten days earlier, I had docked my boat, the *Stella Maris*,*
in West Vancouver after completing a long, solitary journey during which I had explored the shores and waters of the north Pacific Ocean. Now, in late September 1972, as I flew some five hundred miles east of the nearest salt water, it seemed strange that my present endeavor should have occurred to me while I was sheltering from a fierce sea storm.

*See R. D. Lawrence, *Voyage of the Stella* (New York: Holt, Rinehart and Winston, 1982).

It was as though fate had ordained that I should come here by first causing me to shelter the *Stella* in the harbor of Kelsey Bay on Vancouver Island, and then, when I went for a walk in the nearby forests in order to kill time, by allowing me to see a puma as it was crossing a rain-swept clearing.

But when I donned a waterproof suit and set out to walk in the wilderness that lies to the east of Kelsey Bay, I merely intended to stretch my legs on shore and to pass an afternoon in idleness while waiting for the weather to moderate. But after I saw the puma emerge from the forest, pause momentarily, and then go bounding across the clearing, my long-time interest in this animal was immediately revived.

My ocean journey was almost over and I had been casting about without success for some new project with which to occupy myself during the months ahead. Soon after sighting the puma, while I was retracing my route to Kelsey Bay, I began thinking about studying the animal, indulging in more or less idle speculation at first. As I explored the subject, I found myself becoming more and more intrigued by it. By the time I returned to the *Stella*, I had become fully committed. I had previously made up my mind to sell the boat once I arrived at Vancouver, and I now decided that as soon as I had made arrangements for the sale, I would enter some part of the British Columbia wilderness and try to locate a suitable puma habitat. If successful, I would spend autumn, winter, and spring studying an animal that had fascinated me for more than twenty years. This would be quite a challenge, for I knew the mountain lion to be one of the most elusive big-game animals to be found in North America.

On September 16 I ended my voyage in Vancouver. Soon after, I made arrangements for the sale of the boat and drove inland, at first going to Invermere, in the East Kootenay region of the province, an area within the Purcell Mountains about 150 miles north of the Montana border. Here I spent a couple of days camping in the mountains, resting after my nonstop drive from the coast and thinking about the project I was about to undertake. It was September 22 when I settled myself in this wilderness; during the next forty-eight hours, while relaxing completely, I marshaled my knowledge of the puma, concentrating particularly on the range that this animal occupies in Canada's westernmost province.

The mountain lion, also called puma and cougar, is found over some two-thirds of British Columbia as far north as the 56th parallel—though there have been sightings considerably north of that latitude. Three subspecies live in the province: *Felis concolor missoulensis*, *Felis concolor oregonensis*, and *Felis concolor vancouverensis*. Differences among them are not readily observable, since they relate to size and color, which are both extremely variable even among animals of the same subspecies. Indeed, in layman's terms, a mountain lion is a mountain lion is a mountain lion! Members of different subspecies share the same inherent traits, behave in similar ways, and are almost the same size. Only a dead lion can be clearly identified through careful weighing, measuring, and close examination of bone structure.

Missoulensis is the most abundant subspecies in British Columbia, occupying the largest interior mainland range, mostly east of the Cascade Mountains; *oregonensis* is said to inhabit the Coast Mountain region up to the Bella Coola valley, but since the ranges of both subspecies overlap, and since they freely interbreed, it is difficult if not impossible to determine where the range of one type of cat ends and another begins. Although I was marginally interested in these distinctions, I was not unduly preoccupied by them, not being fond of such biological hairsplitting. But inasmuch as *missoulensis* was, at least theoretically, the most plentiful, I was determined to concentrate on this cat.

Having more or less selected a geographic area of operations, I now sought to narrow the choice of territory, for *missoulensis* occupies a range that encompasses about 140,000 square miles. Some of this is given over to civilization, where, of course, there are fewer cats—although they are by no means rare even within the environs of small cities. By the same token, this predominantly solitary animal is never found in large numbers in any part of its habitat, so my task was rather reminiscent of the recipe for hare pie, which begins: *First catch a hare. . . .* Before I could expect to observe the animal, I needed to decide on a piece of country in which to search for it; then I had to make sure that at least one mountain lion was to be found there.

Knowledge of an animal's habits and preferred environment is essential if one hopes to locate it within a large region of wilderness other than by pure chance, so a would-be observer must know where to look. If, for example, one is seeking antelope, the best place to search for it is on quiet sections of the prairies; this makes such a quest relatively simple, for the pronghorn is a highly specialized animal that can only survive in one kind of habitat. Not so the mountain lion. It is an opportunist both in diet and homesite, and is as much at home on the forested slopes as it is in the valleys, cut-over land, and even pastures, provided these are bordered by treed wilderness. But the animal *does* have a distinct preference for rocky and mountainous country where deer, its principal prey, are to be found in reasonable numbers. This knowledge did not help me narrow my search, for most of the province of British Columbia is composed of such ideal puma habitat.

After debating the matter for a time, I concluded that what I really had to do was seek a range that would suit my own needs while fulfilling the puma's requirements. During my first evening in the Purcell Mountains I decided I would prefer a more southerly part of the wilderness, one where the cold would not be so severe and where winter comes a little later and leaves a little earlier. I also preferred an area near enough to some form of civilization to allow me to reach it without an enormous amount of trouble, but which would present obstacles sufficient to deter casual human travelers. This section of wilderness should have valleys and waterways sheltered by mountains, be well timbered, yet offer a variety of open locations where deer and other ungulates would find plenty of winter feed.

Aided by my own knowledge of the province as well as by a general map and a number of large-scale survey maps, I finally settled on the territory that lies within the Big Bend of the Columbia River, a region bounded on the east by the town of Golden, on the west by Revelstoke, and on the north by Mica Creek. Here, within 160 square miles of wilderness, where the only roads are rough trails or waterways and where not even a village exists, was a country that fulfilled my own and the mountain lion's needs. This locale, in the heart of the Selkirk Mountains, is isolated enough to discourage all but the most determined explorers and is far enough away from sources of supply to require that those few who may wish to remain within its boundaries for any length of time must first be prepared to make several arduous canoe journeys into the wilds in order to ferry in food and other essential supplies.

When I ended my two-day stay in the Purcells and returned to civilization, I hired the plane and pilot and set out to search the wilderness, for I actually had to *see* a mountain lion occupying a particular section of range before I could make further plans. After watching the cat from the air as it disappeared, I asked the pilot to fly out of the valley, turn full circle, and return so that we could get an accurate fix on the cougar's location. The valley was relatively flat, rising only gradually almost due north for eight miles. Overlooking it were icefields where mountains closed their ranks rather abruptly. Down the center of the valley ran a little waterway called French Creek, which was fed from a glacier that sprawled close to the peak.

While the plane began to climb, I continued to study the ground, orienting the large-scale topographic map that rested on my knees and noticing for the first time a small clearing that seemed to contain the ruins of a wooden building. Since this little dale was close to the slope the lion had descended, I used it and the old building as a landfall and was in this way able to obtain a good set of bearings. As we were leaving the area, I charted the figures on the map: 51° 39′ 28″ N; 118° 25′ 26″ W. Besides this information, I wrote the words *Base Camp*, for I had already decided to make the valley my headquarters when I returned here by canoe prepared to remain in the wilderness until next year's floodwaters had spent themselves.

Later, in Revelstoke, I checked into a motel and began to make preparations for my expedition. It was September 26. If I hoped to settle myself in the area before the end of the month, I had no time to lose.

Field research of the kind I was planning demands considerable advance preparation over a period of time greater than I had at my disposal. But despite the difficulties, I was fully determined to study the mountain lion I had seen, or at least *try* to study it, for I had no guarantee that I would see it again, or that it would remain in the same territory once I intruded. I viewed such negative considerations briefly, then discarded them in favor of the more positive aspects of the undertaking. First, I had to buy or rent a canoe; and I had to equip myself for a short preliminary stay in the wilderness and then find the French Creek valley.

Studying a map, I saw that a paved highway linked Revelstoke with Mica Creek; it traveled east of the Columbia River. The distance from Revelstoke to the Goldstream River, which I was going to have to use to reach the valley, was about fifty-five miles. A mile northwest from where the Goldstream passes under the highway to empty itself into the Columbia, a gravel road led eastward for two miles toward the river's conclusion, and passed under the main north-south power line above a series of rapids. Beyond the power line there appeared to be a trail leading to the French Creek valley, but this winding, narrow track that would almost certainly be obstructed by second-growth trees was of no use to me. I had to move quickly, transporting a lot of supplies and equipment; the only way to do this would be to travel by canoe up the Goldstream River. I now reasoned that if I could get as far as the power line in my station wagon, it would then be relatively simple to follow the river until I reached its junction with French Creek, then walk the two miles to the little valley where the tumbledown shack was.

That afternoon I went shopping, eventually buying a used sixteen-foot Chestnut canoe that a dealer had stored in the back of his premises; it cost $250, but was cheaper than renting on a long-term basis. When the craft was strapped to the roof of the car, I drove to a supermarket and bought enough food to last three days. (Although I only planned a forty-eight-hour stay in the wilderness, past experience had taught me that it is always wise to take more food than is actually required for the duration of the stay; things can go wrong and cause delays.) I had an early supper, returned to the motel, where I paid the bill for my room, and went to bed early.

I left Revelstoke at five o'clock the next morning and reached the gravel road in the vicinity of the Goldstream a little more than an hour later, finding that though this thoroughfare was dry, it was narrow and bumpy, in places obstructed by fallen branches and saplings that had to be cleared before the car could proceed. Nevertheless, I covered the mile and a half to the power line in twenty minutes and there, hacking a space in the underbrush, I turned the car and parked it.

Before doing anything else, I walked about two hundred yards to stand on the riverbank, noting at once that the water flowed slowly, as I had expected, for in autumn the melt from the high peaks has ceased. But the gurgling of the river as it traveled over the rapids located about a quarter of a mile west of where I stood was clearly audible, perhaps augmented by the rumble of the more distant falls, for there are seven of these interposed along the last mile of the Goldstream.

Once I was satisfied that I could safely begin my canoe journey from this place, I returned to the car, unstrapped the craft, and carried it to the water's edge before returning again to collect food and equipment. It was after 7:00 A.M. when I was at last ready to leave.

This first reconnaissance trip up the Goldstream was of major importance, for if I was to succeed in my project, it was essential that I get to know the waterway and the country that flanked it. Because of this, I paddled slowly and devoted much attention to the route, stopping often to mark some salient point on the survey map, leaving small cairns of rocks in various locations, and charting their positions. While traveling in this way, I had an opportunity to enjoy the journey and to see some of the birds and mammals that lived in the environs of the river. My first sighting was of a female black bear who was accompanied by two yearling cubs—gangly, half-grown youngsters who bolted as soon as they detected my scent. Not so the mother; she stood on her hind legs, her big feet planted securely on the water-lapped shingle as she sought to get a better look at the stranger who had invaded her domain, while sniffing at me and raising her upper lip above the gums to reveal her yellowish and formidable tusks. I had stopped paddling, but as the current began to take the canoe downstream, I plied the blade again. At this, the bear smacked her lips, making an audible slapping sound, which was a sure sign that she was getting nervous. I spoke to her quietly as I continued on my way and she dropped to all fours and backed into the forest.

As I turned a bend in the river a while later, I noticed a bushy-tailed wood rat (*Neotoma cinerea*), or pack rat. This is the thief that invades cabins and tents and steals anything it can carry away, a swift but inordinately clumsy and noisy robber. The rat was just scuttling down the bole of an Engelmann spruce, a bundle of evergreen twigs and needles in its mouth, when I showed up. Dropping its little hoard, it squeaked shrilly, outraged at my presence, then turned around and scurried up the tree. It was about seventeen inches long from nose to somewhat bushy tail, dark gray on the top and white on the underside, its body a deep, tawny color on the cheeks, shoulders, and flanks; its ankles and feet were white. This is the only native rat to be found in Canada, an animal that is most unlike the disease-carrying European varieties. Indeed, it doesn't look ratlike, resembling a squirrel instead. Typically western, its range extends from the extreme western edge of Alberta to the Pacific coast and as far north as the southern Yukon, then southwestward into the United States to New Mexico.

Two beavers also crossed my path, and muskrats were evidently plentiful along the river, for I saw nine of these interesting animals. As I was nearing the junction of creek and river, I also caught a brief glimpse of a red fox, but this little hunter was too wise to be caught in the open. It had evidently heard or scented me before I was near enough to see it clearly.

There were also many birds in this habitat, but I mainly heard them as they foraged for insects or seeds within the shelter of the forest, though I sighted several nuthatches, many chickadees, and five hairy woodpeckers. All of this pleased me, not only because I enjoy living in wilderness areas that contain abundant life forms, but also, and more important, because they indicated that the land offered plenty of food and shelter. Such fertility, coupled with the sightings from the air, suggested that the puma I had come here to find was probably a permanent resident, rather than a cat that was merely passing through as it searched for a suitable home range.

So far the weather had been kind, with blue skies and sunshine and temperatures ranging in the low seventies, although the nights were chilly; yet, sometime before my arrival in this part of British Columbia, snow had fallen on the high peaks, dusting the tops of the mountains. From now on, I knew the snow would creep toward the valley until it eventually covered the lowlands to a depth of two feet or more. Before that time came, I had much to do if I was going to face the winter inside a warm and permanent shelter.

The Goldstream River travels some seventy-five miles from its source to the east of Downie Peak, which is a jagged "tooth" of the Selkirks that overlooks a multitude of glaciers. Between Downie and Mount Sir Sandford, the crowns of most mountains in the area are partially ice-covered the year round, for glaciers and icefields are commonplace here. Some of these permanently frozen areas are small; others, like Sir Sandford, Argentine, and Pyrite glaciers, are large. From such prehistoric accumulations of blue ice, the Goldstream gets most of its annual water supply.

Halfway along the river's course, the valley through which

it runs becomes wider; marshland appears and vegetation changes from alpine to lowland varieties, which furnish browse and grazing for deer and woodland caribou and a variety of smaller plant-eaters. Predators are found here also; timber wolves, coyotes, lynxes, wolverines, and, of course, mountain lions, are the major hunters, but many of the lesser meat-eaters share the domain.

From the junction of French Creek to the point where the Goldstream passes under the power line near the river's terminus, the valley is more than a mile wide. Here the waterway twists and winds through areas of marsh that are interspersed with high land on which the trees grow tall, for in such locations there are abundant moisture, good light, and rich humus to nourish the mixed species that have established themselves on favored sites.

The distance from the power line to the mouth of French Creek is about fifteen miles, but it is difficult to give an exact figure because of the circuitous course that must be followed. Normally I would have expected to complete such a journey in less than three hours, but the many stops that I made delayed me so much that I did not reach French Creek until one-twenty in the afternoon. At this juncture of travel, I found that the creek was navigable for a further half-mile, as far as a place where several tributary streams joined it. Here, a long and fairly wide sandbar had formed and become split by the trundling waters, which had carved a channel through the middle of the bar. I beached the canoe and shouldered the backpack and tent before climbing the shallow bank and noted an old trail that I knew led to the valley of the shack, having seen it from the air.

Overgrown through lack of use, the forest track ran parallel to the slopes, but was still passable, having been kept more or less open by the animals that had used it as they went to and from the lowlands. Nevertheless, when I at last reached the trail's end about a mile away from where I had started, I was glad to stop at the edge of the little dale, for my load weighed seventy-five pounds. For some moments I stood there, looking at the opening and at the surrounding wilderness.

The French Creek valley, as I had already noted during the flight and also confirmed from the map, is actually relatively flat. It is shaped somewhat like a bow, and Engelmann spruces and tall larches are its most numerous trees; they grow in dense copses in some locations, but on more open landscapes they stand alone or in small groups. Nearer to French Creek are willows, alders, poplars, birches, and an assortment of other deciduous trees and shrubs, their numbers and heights varying according to the nature of the soil and the amount of moisture and light available.

The little clearing that faced me can only be described as a valley within a valley, a place slightly above the bottomlands, but of open vista and gently sloping contours. The most prominent ground feature of the alpine meadow was the collapsed building, which the map had identified as the onetime headquarters of a small mine that, like so many of its kind still to be found in British Columbia, was abandoned when its mineral deposits ran out. The entrance to the diggings had caved in, completely filling the wound that man had inflicted on the land during the early years of the present century; the building, a long clapboard structure, had been converted by time and the elements into an irregular pile of boards and studdings. Many of its planks and two-by-four timbers were rotten and useless; others, milled from erosion-resistant cedar, were still sound as they lay in an untidy heap.

To the east and west, mountains clothed in tightly packed spruces rose abruptly, their peaks hidden from view; but within the valley itself, only a scattering of young trees had managed to take hold, perhaps because ancient landslides had covered the bottom with flinty soil and broken granite boulders. Grasses, lichens, and small plants dominated the scene here, giving the place a parklike appearance.

After looking around, I left the trail. I was about to take off my backpack in order to look for a suitable campsite, when a patch of torn-up ground some yards to my left caught my attention. The grasses and old leaves that had recently covered the earth had been mostly scraped off and mounded together in the middle of the terrain, forming a small hill about a foot in diameter at its base, some 8 inches across the top, and about 6 inches high. Around this mound the raked surface covered an area roughly 3 feet long by 2 feet wide. Ridges of grass survived in curving lines of green no more than half an inch wide; the soil itself was striated, as though crudely cultivated by a farm implement. All the asymmetrical lines and ridges etched in the ground radiated from the mound.

On the bare earth within the raked section were a number of large tracks, some clear and full, others only partial imprints, blurred by small stones and grass. The marks had recently been made by an adult mountain lion, a big, tawny cat that had felt the need to empty its bowels, had stopped here to do so, and had then tarried to cover its wastes by scraping up the surface of the land with its powerful front paws, drawing debris backward and forming the mound. After studying the tracks, I bent down to measure them with the small metal tape measure I always carry when doing field research. The impressions left by the cat's forepaws were 5 inches wide and $4^{1}/_{2}$ inches long; those left by the hind feet were $^{1}/_{2}$ inch smaller all around. The size of the tracks suggested that the lion was probably somewhat larger than the majority of its kind found in that region of British Columbia. The majority of adult male pumas leave footprints about 4 inches in width by $3^{3}/_{4}$ inches in length when they are walking unhurriedly; when they are bounding or running, the tracks they leave are wider and longer, because more weight is placed on the feet. Even when strolling casually, all animals leave larger tracks in winter because of the spreading and melting qualities of the snow.

Scratches, as they are called, left by pumas are merely large replicas of those made by the domestic cat when it covers its own wastes, an ancestral habit that has been variously interpreted. Laymen believe that their felines are being clean when they cover their ordure; biologists are of the opinion that the mountain lion makes such scratches in strategic parts of its range in order to tell others of its kind to keep out of the territory it has selected as its own; conversely, when left by a female that has entered the period of estrus, or heat, it is thought that such scratches serve to notify male lions of the female's willingness to accept their attentions.

On the day that I entered the small valley in the Selkirks, I was not at all convinced that such opinions were entirely valid. Indeed, it was to try to find answers to this question and many others that I had committed myself to doing a long and careful study of the puma, and the fact that I had so quickly stumbled over one of the big cat's scratches caused me considerable satisfaction. Knowing that it would have been most unlikely that two adult pumas would be found in such close proximity to each other, I felt pretty sure that the distinctive cat-tracks had been left by the lion I had come to seek.

After studying the tracks for a few moments, I freed myself of my backpack and then bent low over the mound, sniffing it and learning from its pungent aroma that it had probably been made during the previous night, or early that morning. Next I picked up two small sticks and used them to part the mound, working gently so as not to mar the little hill's contents, and discovering two oblong feces. Using the flat rock as a table, I carefully separated the fibers of the specimen, examining the material closely with a powerful magnifying glass and being rewarded by the discovery of a number of deer hairs interspersed with bits of bone and sinew. In addition, there were some shorter, finer hairs that I was unable to identify. I recorded these observations in my notebook, adding that because it takes between two and three days for wastes to pass through the digestive system and arrive at the sigmoid process for discharge, I estimated that the food that had produced the feces had been ingested some seventy-two hours earlier. This was important information, for it suggested that the cat might be hungry by now, unless, of course, it had made a kill more recently. I now became anxious to find a temporary campsite, pitch the tent, and set off on a more thorough examination of the valley and its surroundings, hoping to find additional signs left by the puma.

In a glade to the north of the pile of timbers, right beside a small stream that tumbled down the mountain to empty into French Creek, I set up the tent. Here there were shelter, fresh water, firewood, and plenty of loose rocks with which to make a fireplace. From the doorway of my shelter, I had a clear view of the small valley and the trail mouth.

By four o'clock that afternoon I had scouted within a mile of my base, and was twice rewarded by finding lion tracks— once on the edge of the creek, a place clearly used by the cat in order to drink, the second time near a marmot burrow, where the lion had lain in wait within a clump of ferns and grasses, which had been flattened by the animal's body. I was about to go closer to the burrow when I saw the marmot, a roly-poly individual that probably weighed about 20 pounds. It was sitting as rigid as a guardsman at Buckingham Palace, just on the edge of its bolt hole, its mantle of long silver fur catching the waning light, causing it to look as though a white shawl had been draped over its shoulders. This species (*Marmota caligata akanagana*) has a relatively short period of activity, usually from April to September, after which it hibernates deep within its burrow. Apart from its silvery mantle, it has a black face, a reddish-buff rump, and light underparts, giving this relative of the groundhog a most distinctive appearance. The one that I observed—and that watched me with at least as much interest—was clearly about to settle down for the autumn and winter. Its coat had the sheen of good health and its body rolled with the stored fat that would be used by its system to nourish the animal during its long sleep.

As I was raising the field glasses, the marmot vented one of its piercing and characteristic whistles, mouth closed, making the sound somewhere within its throat; when I didn't move, it ground its teeth audibly, its lower jaw moving from side to side; then, venting one more loud call, it bolted down its hole, no doubt intending to stay there until spring, for it was already somewhat late in the season for this species to be above ground. The puma had obviously wasted its time by stalking this alert quarry, but the signs it had left behind encouraged me to believe that this cat was probably the local resident.

I was inclined to linger over my survey of the French Creek valley and especially to look for further lion signs, but I put aside the temptation, reminding myself that this was but a preliminary reconnaissance principally undertaken to discover whether or not the lion I had sighted from the aircraft was using the area as part of its home range. I was now satisfied that this was the case, and in view of the advancing autumnal season, I deemed it best to cut short my stay in favor of returning to Revelstoke to prepare for the long-term study.

I was going to have to build myself some sort of dwelling to use as a home base, a place where I could sleep comfortably—and *warmly*—during the snowbound months that lay ahead. Also, I was going to have to either construct or find temporary shelters in strategic locations where I could sleep in marginal warmth during emergencies. But before such things could be accomplished, I had to buy all the supplies and then ferry them in by canoe, load by load, as far as French Creek, whence I would have to backpack them along two miles of uphill trail to the valley of the shack. And all of this had to be done quickly, for I knew that within the next few weeks the first winter snowfalls would descend on the valley bottom.

During late afternoon of the second day I walked away from the campsite, leaving the tent there, but taking the empty backpack. From a tree I suspended the leftover dry foods I had brought.

That same evening I again booked a room in the Revelstoke motel before going out to supper. Later I began to make lists of the food and supplies I would need in order to survive until the spring. When I had completed these and double-checked them, I spent several hours thinking about the project and making sketches of the shack that I meant to build in the small valley. There was more than enough cedar lumber available among the ruins of the mine building to construct a dwelling 10 feet long, 8 feet wide, and 6 feet high. The cedar, which offered two-by-four studding for frame and roof rafters, was an unexpected bonanza that would simplify my task while eliminating the need to build a much less secure shack out of logs.

Supplies had to be kept to a minimum because of considerations of weight and space in the canoe, so I was going to have to forgo buying roof insulation. But a trick I had used in the past would serve almost as well: After the roof was completed—I hoped before the first snow fell—I would cut evergreen boughs and pile them on top of the shelter to a height of about eighteen inches, creating a thick mat built in layers and containing many insulating air spaces. When the heavy snows arrived, each new fall would add more insulation atop the cushion of spruce branches, preventing ice buildup from the stove heat.

A medium-sized, light, airtight heater would serve to keep me comfortable in a building of the size I proposed to erect. These stoves are lightweight and easily portable. The chimney, in eighteen-inch lengths, is also light and quickly erected. Having been a journalist, I knew that I could buy a supply of used, thin aluminum printing plates from the local newspaper with which to cover the roof and make it waterproof. I had also decided to buy two small windows of the kind that hinge at the top and open inward from the bottom, and are often used in basements. These would be something of a luxury, but despite their weight and the need to guard the glass against breakage during transportation, I felt that I could not spend an entire winter in a windowless building.

As soon as the stores were open the next morning, I went shopping. My first stop was the local newspaper, where, for a few cents each, I purchased enough used plates to cover both roof and floor and emerged with a neat bundle that weighed a little over 20 pounds. In a nearby hardware store, I bought some $2^1/_2$- and 4-inch nails, a purchase that added 35 pounds to my load. I also bought two rolls of heavy polyethylene plastic sheeting with which to draft-proof the walls. This would be applied on the outside of the studdings, then covered by the wall boards. The windows were 3 feet wide by 18 inches high, each weighing 15 pounds, bringing my total hardware weight to 85 pounds.

Food is always the greatest problem in expeditions of this kind, for it represents the largest bulk and the greatest weight. After years of experience, I knew that my personal daily requirements amounted to between 2 and 3 pounds of nourishment, depending on the amount of energy expended. In this instance, since I planned to stay in the wilderness until the following spring, safety demanded that I estimate my intake at 3 pounds per day for a period of no less than 300 days: 900 pounds! Of this, the bulk would be made up of dry stores, such as beans, rice, dried fruits, powdered milk, nuts, soya beans, raw oatmeal, whole-wheat flour, tea, coffee, and some raw sugar. Except for twenty-four small cans of corned beef, I did not intend to take any meat, but I would pack a good supply of OXO cubes and dehydrated soups containing meat products.

By the middle of that afternoon I had purchased supplies totaling 1,100 pounds, a load that fitted into my station wagon, but caused it to sag somewhat. Since I had left all my camping equipment and tools in the valley of the shack, and had hidden the canoe in the brush on the banks of the Goldstream River, the only other heavy items to add to the load were the airtight heater and stovepipes and the ten books I planned to take.

Before the end of that day I had made arrangements with a garage to store my car and had found a man who owned a four-wheel-drive truck and was willing to carry me and the supplies to the Goldstream for fifty dollars. We arranged to meet at the motel at 6:00 A.M. the next day. When I was ready to leave at the end of my study, I intended to hitchhike my way back to Revelstoke, pick up the station wagon, and retrieve the canoe and camping equipment.

 BEFORE LEAVING EUROPE TO LIVE IN CAN-
ada, my knowledge of the puma had been
gained from books and from studying those
of its numbers that were confined in zoos:
Hagenbeck's in Hamburg, Regent's Park in
London, the zoos in Paris and in Barcelona.
Each had one or more of the cats imprisoned in concrete
cubicles, well fed, usually, but lethargic and sullen. Often
they paced continuously, but they would intermittently squat,
or lie down and stare into space, their yellow eyes cold and
vacant. I noticed that visitors didn't linger in front of the bars
of their cages. There was more excitement to be had in front
of the African lion pens; and the big tigers were always intriguing. But there was something about the puma that fascinated me even then, an aura of wildness and pride that for
me was more majestic than the mystique that surrounded the
African and Asian cats.

Early one summer morning, I went to Regent's Park Zoo
and stationed myself outside the puma's cage, putting my
elbows on the guard rail and settling to watch the tom. He
was lying on his side, immobile, his great front paws stretched
out, showing the thickness of his "wrists" and forelegs. There
were few visitors around at that time of day and none within
the vicinity of the lion's cage. Impulsively I began to talk to
the cat.

Leaning far over the rail, I spoke quietly and was amazed
when the vacant eyes showed interest and turned to look at
me. I continued talking.

Presently, without moving his gaze from mine, the puma
tensed his front legs, his claws extruding slowly as he spread
and arched his toes. Two or three times the dark and shiny
claws emerged and retracted in slow motion before the lion
raised his head and shoulders and placed his paws flat on the
concrete. He began to purr. I could see the quiver in his

throat as he made the deep, husky sound that, except for its volume, was almost exactly like the rumble of contentment made by a domestic cat.

After that day I became a regular visitor, and our friendship developed and ripened during the next two years. At first, however, I was merely intrigued, limiting my visits to two or three times a month, but when I realized that the cat seemed to look forward to my arrival, I went more often.

One evening I reached the zoo late, nearly at closing time. As usual, I leaned over the rail and began to talk to Tom, as I now called him, and he reacted characteristically, purring, striding to the bars, and rubbing himself against them, the end of his black-tipped tail jerking spasmodically from side to side. I always spoke quietly, as much because I do not like loud voices as because in those times I was frankly embarrassed by the prospect of being found conversing with an animal; so I was rather taken aback when I heard another voice immediately behind me.

"You're good for that cat. He's much more interested in what's going on since you started talking to him."

Feeling somewhat foolish, I turned around and encountered a short, gray-haired man dressed in a zoo uniform. He was the head keeper of the feline exhibits, he explained, and he and the other keepers had been concerned about the tom's condition ever since the animal had arrived eighteen months earlier, a gift from a lady in Arizona. The cougar had at first paced endlessly, eating little and losing weight. Later, although its appetite improved, it became listless, moving only occasionally and spending most of its time lying on the concrete, staring vacantly into space. Since I had begun to visit the cat regularly, it appeared to be showing more interest in its surroundings and had begun to scramble on the inclined tree trunks that had been put into its cage.

"It really responds to you! As a matter of fact, you seem

to be the only one that it does respond to," the keeper said.

He explained that the cat was intractable and had to be shut in its sleeping quarters when its cage was being cleaned out. It had even tried to reach out with its paws to attack keepers who walked too close to the bars of its cage. At this news, I was seized by an intense desire to climb over the spectator railings and go up to the tom, but the zoo official was reluctant to accede to my request to do so. After considerable persuasion, however, he agreed to allow me to enter the forbidden space behind the rails, provided I didn't get too close to the puma's bars.

Talking continuously and softly, I walked toward the cage, stopping about two feet in front of the bars. Tom had watched me fixedly while I approached, but he didn't stop purring. Now, as I continued to speak to him, he moved toward the bars and began to rub himself against them, much as a cat rubs against a table leg. The purring increased in tempo.

Behind me, I heard the zoo man speak softly: "I'll be damned!"

Emboldened by the cougar's obvious friendship, I moved closer and reached out with my right hand, aiming for the cat's right shoulder. I felt apprehensive, but I could not restrain my action. Gingerly, as the cougar rubbed himself back and forth, I scratched his hide, an intermittent contact at first. But moments later the cat stopped moving and instead pressed himself against my fingers as I touched him. I became bolder and dug deeper into the tawny hair, scratching gently at the skin and inching forward toward the head. Unable to reach the nearest ear because of one of the bars, I withdrew my hand. The tom looked at me, his eyes friendly. He purred more loudly. As he watched, I reached in and began to scratch his head and he again pressed against my hand. In this way, I spent almost twenty minutes fondling the great golden animal while he purred, twitched his hide in excitement, and continued to press himself against my fingers. The keeper, standing some distance away, was astounded. Only half-jokingly, he afterward asked me if I would like to work at the zoo.

From that evening on, I was always allowed to fondle the puma, who showed in every way that he enjoyed the attention and often would roll over on his back, as close to the bars as he could get, so that I might scratch his chest and stomach. Yet, whenever a keeper approached too near, he would spring to his feet, back away from the bars, and bare his great tusks, a low warning growl issuing from his throat.

In some mysterious way the cat and I had become attuned, but although the zoo men insisted on giving me the credit for this relationship, I have never felt that it was entirely my doing. I had, I suppose, initiated our friendship by talking to the animal while at the same time feeling deeply interested in him; yet it was the cat who responded, who voiced his pleasure, and later trusted me to the point where he allowed my hands to explore his body.

After some time, one of the zoo men told me that Tom knew when I was coming to visit him. He would become excited and start to pace restlessly back and forth in his cage, while looking intently toward the path along which I would arrive. I did not believe this at first, thinking that the keeper was imagining it, but when other zoo officials also noticed the cat's behavior, I had perforce to accept that Tom could, indeed, sense my presence in the vicinity of the zoo.

Unwittingly, the zoo men later confirmed the animal's astonishing perceptions when they started a small betting pool, the winner being the one who could guess my arrival with the greatest accuracy. By now, they had told me that the cat's first signs of expectancy were preceded by a series of low but clear whistles, a characteristic of male cougars that in the wild is usually employed during courting rituals, when a tom replies to the fearsome cries of passion voiced by a female in heat.

Because of these impromptu wagers, I was able to learn that Tom was somehow able to sense my arrival while I was still from eight to eleven minutes away from his cage, which was more or less the amount of time it took me to enter the zoo and to walk directly to his concrete-and-iron prison. The subject of extrasensory perception was not as widely discussed or studied then as it is today, yet I realized that my arrival outside the zoo gates usually caused me to direct my thinking toward the cat, my mind remaining centered on Tom until I stood before him. I did not, however, pursue the matter to any great extent, being more intrigued by my friendship with the cat than by the seemingly unfathomable mysteries of ESP. Thus I missed a perfect opportunity to do some serious research on a subject that was to interest me greatly many years later.

I could not, of course, visit Tom every day, but I did so as frequently as I could, usually being able to spend an hour or two in his company, five or six times a month. Thus our friendship ripened and flowered into an almost uncanny communion between two quite disparate beings. I could do just about anything with Tom—even, I felt, enter his cage. But the zoo people would never allow me to do this, despite all my entreaties and vows to sign any paper they wished clearing the zoo of responsibility in the unlikely event that Tom might decide to attack me.

For almost two years Tom and I continued to commune at every opportunity; then, after I had been away on assignment in Algeria for nine weeks, I returned to find his cage empty. On inquiring as to his whereabouts, I was told that Tom had developed a tumor on his shoulder and had died. Saddened by the loss of my friend and feeling guilty because I had been away for so long, I left the zoo, never to return again.

After arriving in Canada in 1954 and during the years while I was homesteading in northern Ontario, I had not devoted much thought to the mountain lion because I was engrossed with the many other animals that I was studying in the spruce forests that surrounded my property. But after I gave up the homestead and left with my dog, Yukon, to explore more of the northern wilderness, I had another encounter with a puma that, though brief, made a great impression on me.

In 1959 I had taken a job with a small newspaper in British Columbia because I needed money to buy freedom for myself and Yukon, so that we could again go into the wilderness for an extended length of time. We were then living in a cabin on a bit of forested property that I had rented a few miles away from the town where I worked.

Earlier, I had met a stockman when he came to the office to place an advertisement in the newspaper. He had realized that I was a newcomer and, as is common in small communities, asked me about myself. I told him about my interest in nature. We talked for only a few minutes, and when he left, I didn't think any more about the casual meeting, perhaps because I had not been very impressed with the man.

So I was surprised when he called me at the office one evening (I worked nights on a morning newspaper) and invited me to go with him next morning to hunt a puma that had "slaughtered" his stock. Unthinkingly I accepted, and we arranged that he would collect me in his pickup truck early the following morning, bringing his two redbone hounds, specially trained to go after mountain lions. This meant that Yukon had to remain at home, which was something he didn't like, but it was unavoidable, for he would not tolerate any dog's leadership but his own and would, therefore, spend the first several minutes thrashing the hounds. (Yukon weighed 120 pounds and was tough enough to bring down a moose.)

Accompanied by my dog's protesting howls and assailed by the deep, hysterical baying of the two redbones, I climbed the hill from cabin to roadway to greet the stockman. There he introduced me to his dogs, and as I was climbing into the truck, he said he was really looking forward to killing the lion.

"I always gut-shoot the bastards! Make 'em suffer some. Boy, these hounds sure love to hassle a gut-shot cat!"

I was about to say some unpleasant things and forgo the hunt, but was stopped by the vision of the cat's agonizing, tormented death. So I kept quiet. I had intended to bring a camera but no gun, not wishing to actually take part in the killing, but now I pretended to have forgotten my Lee Enfield and went back to get it, leaving the photographic equipment behind.

My companion didn't notice the absence of my camera until we were approaching the place where the hounds had been pulled off the cat's scent the evening before, it being too late then to continue tracking the cougar.

"Why'd you leave the camera behind? I was hopin' you'd get some shots of the kill," he remonstrated. I suddenly understood his reason for asking me along. Knowing I was both a journalist and an editor of the newspaper, he had visions of a nice spread, no doubt seeing himself posed, one foot on the dead cat, rifle butt resting on the inert carcass.

Since I too was scheming now, I said that we could get pictures afterward, providing the puma was indeed located and killed. I didn't listen to his reply, being busy with my own thoughts, for I had decided to shoot the puma in the head, even, I hoped, as this coarse man was squeezing his own trigger, or immediately afterward. That lion was not going to be left to die from a stomach wound while my companion got his jollies as he watched!

On our way to the place where the hounds had been pulled off the scent yesterday, the stockman detoured to show me the remains of a ewe, which had been killed by one powerful blow that had smashed the skull and broken the neck. The lion had eaten some of the intestines, the lungs, the liver, and part of one haunch, then it had covered the carcass with leaves and forest duff.

The hounds, already frantic to begin the chase, nosed at the dead sheep, whining excitedly; then, as the stockman drew them away from the remains, they set their noses to the trail, pulling eagerly.

For the next two hours the dogs struggled against their leads while we worked deeper and deeper into the wilderness. Up to this point, I held one dog in leash, their owner holding the other, no inconsiderable task in view of the nature of the terrain. Presently, beyond the point where the man had called off the hunt yesterday, he slipped the leads.

The dogs shot away, baying intermittently, tails curved high over their backs. We had to run now to try to keep as close to them as possible, because if the lion should turn at bay, the dogs would charge it and would almost certainly be torn to pieces unless we were on hand to end the matter.

I was lighter and fitter than my companion, and I soon began to outdistance him. At first I would stop and wait for him to catch up, then run again, but as the sound of the dogs grew more distant, I kept on going. Later, in the excitement of the chase, I quickly forgot about the stockman.

When the baying of the hounds became deeper and faster, I knew that they were now hot on the cat's trail and that the quarry was doing a lot of twisting and turning. This allowed me to take shortcuts, although I did a lot of zigzagging myself as the chase went this way and that. Quite suddenly the sounds of pursuit began to come toward me. The cat had turned. I stopped, panting and trying to listen above the sound of my rasping breath. I judged that the hounds were about half a mile ahead now, which meant that the puma might be a lot closer. Before I had time to react, the baying became higher, faster; seconds later I realized that it was issuing from a static location. The cougar had climbed a tree.

I called to my companion, listened for his reply, but none came. The voices of the dogs told me that they were becoming hysterical; I could visualize them jumping and pawing at the foot of the tree in which the cat had taken shelter, and I waited no longer. Even now, if I could reach the hounds in time, I might be able to restrain them and give the puma a chance to escape without the knowledge of the stockman. Five minutes later I found myself jogging through country that was sparsely covered by spruces, larches, and a number of young birches. It was sloping land, not too steep, but littered with old stumps, evidence of the logging that had created the clearing. Ground cover was bushy. Thimbleberries grew profusely, the bright red fruit adding a Christmasy look to the landscape, while thick clumps of prickly broom, each decorated by pealike seed pods, continuously interposed themselves in my path, their spiky branches clutching at my clothing as though they would prevent me from going any farther. The most predominant shrub in the area seemed to be mountain Labrador tea (*Ledum groenlandicum*), which grew waist-high in company with swamp laurel, for the foothills were quite moist here, even boggy in places.

It took about ten minutes to reach the place where the hounds were frantically and fruitlessly trying to climb a tall, thick ponderosa pine that, typically, was clear of branches for much of its height. The tree was partly dead, only the uppermost limbs retaining green needles. The puma was precariously perched about fifty feet from the ground, its hind feet resting on a branch that could not have been more than two or three inches thick, its forefeet gripping a similar branch about two feet lower down and to the left of the upper perch. The cat was looking down, mouth agape as it snarled, its large and gleaming fangs formidable weapons within the pink cavity of its mouth.

I stopped about fifty feet away, catching my breath while looking at the great, tawny animal and at the baying, leaping dogs tormenting it. I raised the field glasses. The lion's unsheathed front claws were working at the branch and dislodging small pieces of bark that fell on top of the dogs. The cat's open jaws were soaked by saliva, some of which dripped from its lower lip. Its eyes were ablaze with rage. As my breathing became more normal, I heard the cougar's snarls, low and rumbling and full of menace.

I lowered the glasses and strode to the foot of the tree, trying to pull the dogs away. At my approach, the cat's snarls became louder and were repeated more rapidly. I could not secure the hysterical hounds, and when a shower of loose bark fell from above, I backed away until I could again see the unfortunate cat. Now it had bunched its feet on the lower limb, an impossible perch on which it rocked precariously; its furious protests grew louder and I now heard a husky, spitting sound. Soon the lion was going to come down the tree. If it did, the dogs would charge and almost certainly be mauled, perhaps killed.

I hesitated, wanting desperately not to kill that magnificent, persecuted animal. But neither did I propose to allow it to be gut-shot.

I heard a call. It was the stockman's voice. He was still some distance away, but he would nevertheless arrive at the tree within minutes. The cat heard the voice also. It scrambled downward, paused on two of the lower branches, snarling loudly, then began to move again. Once more I heard my companion's voice; he was appreciably closer.

About twenty feet below its original perch, the puma paused again, its gaze focused on the dogs as it prepared itself for the last frantic scramble. Now it raised its head and looked at me, its snarls undiminished, its yellow eyes filled with fury. I had to shoot.

With deep regret and a feeling of nausea, I aimed the rifle, taking a fine bead at a point exactly between the eyes, allowing for wind deflection, height, and range.

The sharp crack of the Lee Enfield was still ringing in my ears when the lion's body hit the ground with a thump, rolled once under its own momentum, and fell over the lip of a ravine that I had not noticed earlier.

During the fractional interval that occurred between the sound of the shot and the noise made by the cat's body as it hit the ground, I seemed to have all the time needed to observe the results of the shot.

For a split second, even as I was squeezing the trigger, that cat's eyes met mine, then I saw its head jerk sideways and upward. I must have opened my left eye at that instant, for I suddenly saw more of the puma, its front legs, its claws folding back reflexively; finally, a heartbeat before it fell, I saw the life going out of the big, sleek body. In an infinitesimally small fraction of time I had killed a vibrantly alive, magnificent animal and reduced it to a slack bundle of meat and hair. I cried.

I was vaguely aware that the dogs had dashed to the edge of the crevasse and were dancing along it; I heard the stockman's footsteps behind me, then his voice.

"*Son of a bitch!* That was a head shot, you *bastard!*"

I didn't turn around. Anger flooded my entire being, a rage such as I had not experienced for a long, long time. I wiped my eyes and cheeks as I tried to control myself, because suddenly I wanted to turn and shoot this gross, brutal human.

It was fortunate that the stockman did not again speak until he had marched to the lip of the drop and stared at the cat's body for some moments. When he turned around again, his eyes widened when he saw my face.

"You all right?"

I nodded in reply. Then I forced myself to go and look at the puma's body. It lay about thirty feet down, sprawled over some rocks. Blood ran from its nostrils and mouth and from the bullet's exit hole at the back of the skull, an obscene crater about three inches in diameter. I stared longer than was necessary, wanting to fix that image on my mind, to remember it always. As I looked, the stockman was explaining that he had tripped and had called out to me, but I had not waited for him. His manner was obnoxious, but I forced myself to ignore him as I told myself again and again that I had at least killed the lion instantly. I also took some satisfaction in having prevented my unpleasant companion from executing the cat in his brutal fashion.

It was necessary to recover the puma's body because the hunt had to be reported to local wildlife officials, and since my companion was too heavy and still too winded to make the descent, I went down while he prepared the small block and tackle he had brought in his haversack for just such a situation. When I reached the dead animal about three-quarters of an hour after it had been shot, I noticed for the first time that it was a pregnant female.

As far as I could determine by her condition, the puma had been within twenty-four hours of giving birth when the hunt began. Her belly was distended by the young she carried, her dugs were heavy with milk, driblets of which had even leaked out during the fall. When I opened her, I found that one of her two dead kittens had moved toward the cervix; the amniotic fluid had already been released, evidently during the chase, judging by the sticky wetness along the insides of her back legs and thighs.

No doubt the exertion of the chase had accelerated the birth process, but the size and weight of the kittens tended to confirm my first estimate. Later, when I weighed them carefully, the little male was 14.5 ounces and his sister 13.0 ounces. The tom measured 12 inches from the tip of his tiny nose to the end of his short tail. (Puma kittens do not resemble adults at first. In addition to the difference in the length of tail, their coats are black-spotted and they also bear dark rings around their tails.) The little cat was half an inch shorter than her brother.

Our return trip was a silent and macabre affair. The kittens were in my packsack, the mother's paws were lashed to a pole that suspended her from our shoulders as we silently picked our way while the hounds pranced around the swaying carcass.

I kept the kittens over the stockman's protests, and that evening I buried them under a cherry tree on the property I rented, a little sentimental gesture overseen by Yukon, who, of course, didn't understand what I was doing, but, eminently able to read my moods, sat stoically a little distance away, watching attentively, his ears erect. Afterward he came and put both paws on my shoulders and slurped his great tongue over my face.

Later that evening, before going to work, I sat quietly with Yukon at my feet, thinking about the day's events. I was tired, having had no sleep for more than twenty-four hours, and I felt distressed, even guilty, at first. Later I became angry again. It was as though I had been cheated out of something precious that belonged to me; and I felt that my species was unutterably selfish and ruthless, a destructive force that has been devastating the wilderness of North America for more than four hundred years.

Before I left for the newsroom, I had made up my mind that someday I was going to really get to know the puma.

ON SEPTEMBER 30, 1972, THE MAN I HAD hired to take me from Revelstoke to the Goldstream River helped me unload my supplies when we arrived at the power line. This was the point at which my canoe journey was to start, and marked the true beginning of my search for the puma. When all my gear was off the truck, I paused to watch my rather taciturn chauffeur as he drove away. Then I began sorting my mountainous equipment.

I was going to have to make three trips to French Creek, so I separated the stores into three manageable piles. First to go would be tempting foods, items that attract bears and wolverines as well as squirrels and other rodents; these included rice, flour, oatmeal, fruit, sugar, beans, and similar comestibles. First I carried the bulkier things, contained in fifty-pound sacks, to the water's edge and placed them in the center section of the canoe; smaller bags and packages I transported in two packsacks. This left an assortment of cardboard boxes that were stowed on top of the relatively flat and soft cargo. When the first load was ready, I climbed aboard and pushed the canoe away from the bank. I began paddling upriver, using the steady, rhythmic long-distance pace I had developed years earlier, a moderate speed requiring a minimum of exertion.

At the end of French Creek the stores had to be backpacked, load by load, over the two-mile trail to the valley of the shack, a seemingly endless and fatiguing chore. As I carried each packful into the campsite, I immediately suspended the food from nearby deciduous trees, high enough off the ground to thwart bears and wolverines, and far enough from adjacent branches to discourage squirrels and pack rats. It took almost six hours to pack the canoe, cover the distance from river to sandbar, unload, and carry the goods to the valley; another half-hour was used suspending the food. I knew that the return journey, assisted by the current, would take slightly less than two hours, making an average of eight and a half hours per round trip, not counting rest stops.

At the end of the second journey I paused for a late supper and a rest, already feeling the effects of all that paddling, fetching, and carrying. Afterward, sitting before the fire, enjoying my first pipe of the day, and waiting for the moon to emerge, I considered leaving the last load, consisting of hardware, until the morning; but I discarded the notion, worried that some human predator might happen upon the goods and steal them or, even more likely, vandalize them, for the power line crossed the river at a point within walking distance of the highway. Perhaps the chances of either contingency's occurring were small, but it seemed foolish to take the risk. Ten years earlier, I knew, the equipment could have been left on the riverbank for a month and would have been found intact; but for more than a decade now, thieves and vandals have increased to epidemic proportions even in the hinterlands.

When the big gibbous moon emerged, I reluctantly set out again, depressed by the knowledge that at least eight hours of hard labor still awaited me that night. But this bad mood was soon dispelled.

I have seen much wilderness during the last twenty-eight years, and it has shown me many rare and wonderful things, but if I had to choose one single experience as being the most unforgettable, I think I would have to select that which was afforded me during the early hours of October 1.

It was almost 2:00 A.M. when I paddled away from the power line, carrying the last and lightest load while the brilliant moon hung over the valley, its fulgent display shining through the trees to create long shadows and coax dazzling flashes from the surface of the moving water. This chiaroscuro effect was at first disturbing, for it limited vision and caused me to paddle slowly and cautiously, but later, as I neared that part of the valley where it becomes wider and where the trees are not so tall and grow in isolated clusters, the light became consistently reliable and even more enchanting.

The river was now a serpentine ribbon of molten silver that was flat and mirrorlike in the distance and ebulliently spangled nearby, especially at the bow of the canoe and within the tiny whirlpools left by each stroke of the paddle. Coming downstream, and again after I turned around, I had heard the sounds made by a number of forest animals, but I had seen none of them because of the penumbra of the evergreens; I had also heard fish jumping, and several of the splashes were loud enough to have been rainbow trout. Now, as I neared a marshy pond where during my first trip I had noted a large beaver lodge, a great horned owl called, its five deep notes drawing my gaze to the bird itself. It was sitting tall on a branch midway up a dead aspen, its feathered horns erect and in silhouette, the rest of its body a dark outline against the background of blue-green sky. The pond itself, located about one hundred yards from a deep bend in the river, was agleam with moonlight and this created extravagant, surrealistic shadows in those places where dead trees, catkins, and reeds interrupted the moonbeams. The owl called again. Its voice was immediately followed by the splash of a beaver, the loud and sudden slap reverberating through the wilderness.

Fully relaxed for the first time since arriving the morning before, I began to observe my surroundings attentively; and I listened. A breeze from the south caused the trees to gossip; the canoe and the river whispered softly; droplets that fell from the paddle blade bubbled as they reentered the water.

I slowed my pace and allowed my senses to work independently; eyes, ears, and nose each gathered different information, yet their messages were instantly coordinated by the mind and as quickly converted into a whole canvas, a near-mystical image of the silvery wilderness.

In this way I covered seven or eight miles and was about two-thirds of the way to the creek when the sounds made by several large animals moving through the nearby forest caused me to concentrate on the left bank of the Goldstream. Moments later I turned the canoe off course, nudging the bow onto the right bank. I was certain that a number of wolves were traveling in my direction, no doubt intent on slaking their thirst. The noise made by the pack became louder and I guessed that the wolves were running now, eager to taste the water they had scented.

I had just eased myself into a more comfortable position so that I could watch while remaining immobile, when seven large, magnificent wolves stepped through the screening trees and stopped in unison just outside the lee of the guardian shadows, legs stiff, heads erect, ears and nostrils inspecting the night. The river here is less than one hundred yards wide; the pack stood about fifty yards from the water, the hind-quarters of some animals concealed by the shadows.

Observing the seven heads and chests through the glasses, I had no difficulty picking out the leader, the Alpha wolf, a big gray male that stood taller than all the rest and was half a length in front of his subordinates. As I watched, I counted the seconds slowly and silently; when I reached twenty-two, the lead wolf moved forward, followed in single file by the others; soon they were concealed by the tall meadow grasses, but the moving tops of these browning plants allowed me to follow the pack's progress.

Again I counted, this time reaching thirty-eight at the

moment that the big male came out of the walled sedges about fifty yards south of my position, but, of course, on the opposite bank. The Alpha wolf stood alone, his pack remaining concealed. He checked the night once more, but the small breeze blew toward me, offering him no man-scent. The big wolf turned his head and looked behind, directly into the screen of plants; he made no sound, but even before he turned to face front again, the rest of his pack obeyed his gesture of command and emerged. One by one they settled themselves at the water's edge, front paws in the stream, muzzles lowered. The sound of seven lapping tongues reached me clearly.

The leader was nearest to me. I altered the angle of the glasses, aiming downward so I could see the water and the lupine head. As the wolves drank, the water around their feet and below their active tongues lost its burnished appearance, but now liquid pearls fell from each partly opened mouth. I felt as though I myself were kneeling beside the wild dogs at the stream—my mouth felt suddenly dry.

The countenance of the Alpha male reminded me poignantly of my old friend and trail companion, Yukon,* although his face had been whiter than the charcoal visage of the wolf. While I watched, the male raised his broad and handsome head and looked upriver, a momentary, casual glance that allowed me to notice a round white mark— somewhat larger than a quarter—on his left ear, near the tip. White Spot, I thought, naming him instantly and hoping that this was the same pack I had seen from the air. If it was, I felt that we would assuredly meet again during the months ahead.

The wolves spent several minutes at the water's edge, drink-

---

*See R. D. Lawrence, *The North Runner* (New York: Holt, Rinehart and Winston, 1979).

ing avidly at first, then raising their heads and looking around, much as the leader had, then lapping again, but more slowly and for a shorter time, ears attentive and nostrils flared slightly to taste the night scents.

White Spot became suddenly alert, his whole body stiffening, his eyes turned in my direction. For a split second I saw the moonlight reflected in his yellow pupils. Then he moved away from the water. But he left unhurriedly, even pausing briefly to nuzzle a smaller wolf that had stood beside him throughout, probably his mate, the Alpha female of the pack. Together they led the way back into the shelter of the grasses and sedges, departing as they arrived, with a swishing sound as they traveled within the drying screen of plants, and a faint squelching as their big paws broke through the old and matted vegetation and sank into the oozy bottom. I waited until the noise of their departure ceased before I resumed my journey.

I was now close to French Creek, canoeing along the most winding section of the Goldstream, which travels in a series of tight U-turns here, reminiscent of the marks left by an earthworm as it slithers over wet soil. The valley at this point is wide and relatively clear of tall trees, but soon afterward the predominantly grass-sedge cover changes to an area of marsh, where catkins, rushes, and bog willows grow in profusion, reducing visibility to mere yards on all sides. Twice while coursing through this section, I heard black bears. The first one must have been close to the riverbank and was suddenly startled by my arrival, for it grunted audibly and dashed noisily back toward the forest. Here, too, a number of late ducks quacked sleepily from the shelter of rushes, their voices signaling that they had detected my presence. Fish continued to rise, sometimes jumping, more often nosing at the surface to leave quickly-dispersed rings of bright water. The moon had traveled some distance away from the valley, its yellow light now reaching the Goldstream at a more shallow angle and making longer shadows. A plethora of green and winking stars filled the cloudless firmament.

So pleasant was this journey that I was almost sorry when I at last nosed the canoe on the French Creek sandbar. But the task of unloading and then carrying the stores to base camp soon reminded me of human frailty. It had been a *long* day.

It was after eleven o'clock when I awakened the next morning feeling stiff and sluggish and tempted to remain in bed, but at the same time eager to begin construction of my winter shelter. Later, sitting in front of the campfire eating brunch while contemplating the mound of boards, I became so anxious to start work that I forgot about my aching muscles. For the remainder of that day, and for the next five, I worked steadily, stopping only when the light was too dim to work by. First I sorted and piled up the wall and roof boards, pulling out or breaking off the old and rusty nails; next I worked on the two-by-four studdings. When I had enough two-by-fours, I began sawing and hammering, constructing the floor, then the wall frames, rafters, and wall boards, and finally putting on the roof. During late afternoon of the sixth day, I sheeted the roof and floor with the aluminum plates and hung the two windows. Belatedly I realized that I had forgotten to buy hinges for the door! This oversight was corrected when I cut the tumpline off the older of my two packsacks, a thick, wide leather band worn around the forehead to help support the backload. Cutting this into four equal lengths, I nailed each piece to the door I had fashioned from a number of two-by-six-inch spruce boards found among the top layers of the collapsed building. Although much of the wood had rotted, there was enough sound stock left to make the door. Cedar would have been easier to work with, and lighter, but door-ways and windows are always the weakest part of any wilderness building, usually the first to be attacked by bears and wolverines drawn to investigate the scent of food coming from inside the shelter. For this reason, after the door was hung, I made stout shutters for the outsides of the windows. The covers were removable, the bottom of each supported by a two-by-four nailed into the studdings.

Into brackets fixed securely to the wall on both sides of the windows I slid strong wooden bars that were held in place by four-inch nails fitted into previously drilled holes; these could easily be put in or pulled out when the shutters were placed in position or taken down, but no amount of buffeting by bear or wolverine could dislodge them. As an added precaution, the shutters, the bars, and the door were studded with nails driven through from behind so that they would protrude on the outside, the three-eighths-inch-long points not capable of seriously hurting an animal, but sufficient to deter such intruders as soon as their paws came into contact with the sharp spikes.

Such precautions might seem overly elaborate to those who have not had personal experience with bears and especially with wolverines. The fact is that these animals are enormously strong and quite capable of broaching the windows and doors of wilderness cabins when attracted by the scent of food, or even by the smell of old cooking coming from an unoccupied dwelling. The bears, asleep in their dens from late autumn until early spring, are not a problem in winter; the wolverine prowls hungrily the year round and is, in any event, more inclined than the bear to rip out a cabin window or door.

It is difficult to accept that an animal not much longer than 3 feet and rarely heavier than 45 pounds can cause so much havoc, but those who have been victims of the wolverine will readily confirm its legendary strength and destructiveness. And as if it were not enough that the creature has wrecked one's cabin and eaten one's food, this impudent marauder will then spray everything with an evil-smelling musk, a nauseous repellent manufactured and stored in twin anal glands. This trait, however, is not motivated by pure malice; rather, it is aimed at protecting from other predators any food that the wolverine cannot devour at one sitting, for this relative of the weasel is always on the go and has an exceptionally high metabolic rate that can only be sustained by frequent intakes of meat. Gorged, the wolverine will have a short nap during which its powerful digestive juices dispose of the feast; then it rises, stretches, belches once or twice, and ambles away in quest of its leftovers or, if none of these are available, in search of new prey or carrion, for *Gulo luscus* is not fussy.

Grizzly bears and black bears, I knew, were present in fair numbers in the Goldstream area, but I had not as yet seen wolverine signs. It was, however, almost mathematically certain that there would be at least one of these animals living in the French Creek valley, just as the river valley would have its own population.

I had not expected to see any wildlife during the days that I spent sawing and hammering, but despite the racket I was making, I was regularly visited by the same pack rat, a male that was quick to take the bits of food I left for him near the campfire. And on the second day, as I was preparing a lunch of bannock bread and honey, three gray jays came to join me, from then on becoming regular visitors.

These jays are the greatest panhandlers of the wild—cheeky, daring, and utterly friendly and seemingly equipped with some sort of radar, judging from their ability to arrive at the campfire whenever a woodsman is about to eat. My three companions were to become so tame that they took to following me through the wilderness, often hitching a ride on my head or shoulders and forever ready to share the trail rations (a mixture of raw oatmeal, raisins, and nuts) that I always carry in a bag suspended from my belt.

A week after I began work on the cabin, I completed the building, struck the tent, and moved in. By noon I was settled, and for the remainder of the day I rested in camp attended by my three friends. By now the feathered trio had so endeared themselves to me that I selected appropriate names for them. The smallest, almost certainly a female, I called Wisa, the next I named Ked, and the largest, whom I took to be a male, I christened Jak.* By themselves these names are meaningless, but together they become Wisakedjak, an Algonkian word used to describe a mythical being who is able to assume any shape that he wishes. Akin to the European goblins, this spiritual entity was the principal character in many of the myths of the North American aboriginal tribes, each of which knew it under a different name. Generically, the supernatural sprite was known as the Trickster. The Algonkians called him Nanibush, Glooscap, or Wisakedjak; the Plains Indians knew him as the Old Man; the British Columbia Salish called him Coyote; some of the Pacific Coast native Americans saw him as the raven; others, like the Haisla, referred to him as Weegit. But always the myths were based upon the spirit's love of pranks, some of which were humorous, others distinctly vicious and gruesome.

When I first met the gray jay while homesteading in northern Ontario, I learned that he was called whiskeyjack, a widely applied name in both Canada and the United States,

*There is no difference in the plumage of male and female jays, but the size and behavior patterns can sometimes help to distinguish the sexes.

but when I tried to discover the reason for this title, I drew a blank. After my initial failures, I wondered if the jay had a fondness for whiskey, so I soaked bread in scotch and offered it to two particularly tame individuals. That 1958 experiment failed utterly! Both birds refused the adulterated bait as soon as they got within sniffing distance of its pungent aroma. I tried rye whiskey next, and obtained the same results. At last I gave up, deciding·that some fanciful English immigrant in either Canada or the United States had christened the bird for reasons lost in time.

Many years later I learned that the Algonkian term for the Trickster was also applied to the jay. The connection between Wisakedjak and whiskeyjack became immediately obvious. It is easy to understand why the Algonkians saw the jay as an alter-ego of their supernatural goblin. The bird has many voices; it is bold, yet mysterious when it wants to be; it nests and hatches its young in January, during below-zero temperatures; its intricately made nest is fashioned so as to fit exactly the contours of the hatching female's body in order to protect the eggs and later the young. And the nest is so well concealed that only long and painstaking observation of the birds will reveal its whereabouts. Lastly, young gray jays are dark in color, almost black, not resembling their parents in the least.

People who have lived or camped in the northern parts of this continent know that this fluffy, large bird (up to 13 inches long) is apt to drop in for dinner uninvited. Some people go so far as to claim that it will fly down and steal frying bacon right out of the pan, but I cannot confirm this story and am inclined to discredit it, having too much regard for the jay's intelligence to believe that it would willingly risk burns from the fire and the hot fat in the pan.

However this may be, on the afternoon of October 7, while I was tidying up my headquarters and making a wooden bunk, a table, two chairs, and a few other items of crude but serviceable furniture, Wisa, Ked, and Jak were constant companions who boldly entered the doorway of my little cabin. They were already landing on my open palm to take bits of food, occasionally pausing on this perch to eat a morsel or two. But more often they took as much as their beaks could hold and flew away to hide the spoils in the forest, a practice in which all jays indulge. By evening, as I finished supper, the trio had disappeared and I was left alone to consider my immediate plans.

The next morning, rising at dawn, I breakfasted with the jays and then set out to begin a detailed survey of the mountain lion's range. I started along the west bank of French Creek and, when I reached the end of the valley, returned by the same route, now going slowly and marking in my notebook those locations where I had found puma signs. In seven different places I noted lion tracks; I also found four additional scratches, but all were old. About midway along the course of the valley, near an area of marsh, I found the remains of an old kill. The cat had pulled down a deer somewhere in the vicinity and dragged the carcass into a patch of densely growing willows, where it ate in comfortable concealment. All that was left of the prey were some of its bones, two hooves, and the partly gnawed skull, which revealed that the victim had been a doe, for there were no antler cups on the headbone. I judged that the animal had been killed two to three weeks earlier, and that the remains not consumed by the lion had been cleaned up by lesser meat-eaters and scavengers.

Near the place where the deer was killed, a small creek tumbled downward through a fairly steep gulch. Consulting the survey map, I noted that this was Graham Creek and that its source was located near the top of a 6,800-foot peak that rose just west of the valley of the shack. In the dry silt at creek-edge I found several lion prints, indistinct because of the material in which they had been made, but recent, for some of the partial impressions had clear edges and showed no signs of erosion. On my way home I determined to explore this narrow cut the next day.

On October 9, again leaving early, I walked to the mouth of Graham Creek, located some five miles north of my base. There I followed the nearly dry watercourse until, about two miles from the French Creek valley, I paused to examine a small patch of snow that should not have been where it was. Located about a thousand feet below the white mantle that covered the mountain's peak, it clung to a steep outcrop of dark granite and was surrounded by mosses and clumps of drying grass. The day was clear and sunny, the bottomland temperature hovering in the low seventies, prolonging a spell of Indian summer that had favored the Selkirks for the last eight days. Thus, the indistinct, milky overlay was unusual enough to cause me to raise the field glasses.

As I was doing so, the snowy bulge moved, changing its outline and revealing itself as a mountain goat that had been standing, facing upslope. Continuing to alter its stance, the goat turned smoothly until it stood broadside, leaning against the pull of gravity and staring absently at the nearby trees, the sun eliciting a shoe-polish shine from its backward-curving black horns. From a distance of about four hundred yards, through the glasses, every detail of the animal became clear, even the rather astonished expression on its face. Such a countenance is normal in the species, resulting from the set of its features; the big, pointed ears appear to jut out at right angles to the face; the rounded forehead ends somewhat abruptly at a pair of slanted, solemn, and staring eyes; while the long and Roman nose also seems to end too suddenly, as though Creation had intended a more slender muzzle but had been interrupted, ending its work hurriedly by sculpting a pair of T-shaped black nostrils above an economical slash of ebon lips. From the curve of the throat to the rather knobby chin, a double beard of coarse, long white hair further adds to the perplexed appearance. But this is only an illusion. In reality, *Oreamnos americanus* is rarely surprised by anything. An animal of sharp wits, it is as agile as a squirrel and more surefooted than a mule, an expert alpinist that can disappear almost magically even when traveling over the most barren, steep, and inhospitable terrain.

As if to prove the keenness of its faculties, the goat turned its head and stared at me, clearly aware of my presence. Yet I was almost fully screened by trees, standing at least 1,200 feet away, on the other side of a narrow ravine at the bottom of which the slow waters of an autumnal creek ran over round and well-washed stones. The gentle trickling mingled with the sound of the wind in the treetops, producing an almost continuous muted background music that masked whatever slight noise I might have made. But the goat had found me. Yet it was not especially alarmed, just aware; it kept staring into the glasses for some moments, then raised itself on its hind legs, from this stance lowering its head slightly so as to keep me in view. No other animal can accomplish such a feat on precipitous granite, but that billy made it look easy and, as an encore, actually turned to face uphill, pirouetting on its black, cup-shaped hoofs like some lumpy but eminently graceful ballerina. When it had turned completely around, it dropped back to all fours and trotted over the treacherous incline as confidently as a human might stroll along a side-walk, and moments later was swallowed by the trees. This was the only mammal I had seen during almost three hours of walking. Once I had passed the location where I had found the feline footprints the day before, there was no sign of the lion, so I soon turned back. I occupied the remaining daylight hours by exploring the east bank of French Creek from a point opposite Graham Creek, and then returned to my base. This afternoon journey yielded encouraging results.

I had crossed French Creek by wading through the marshlands that flank it and, while doing so, discovered a series of small lakes nestled within the marsh country, or on its borders. These little lagoons were obviously spring-fed, their waters crystal-clear, the overflow entering the marshlands and eventually emptying into French Creek. Here, along a section of land that measured about four square miles, I found numerous puma signs, including the remains of a dead mule deer that were so fresh I felt the animal had been downed either last night or early this morning. The carcass was half-eaten, both haunches, the stomach, and intestines having been consumed. The ribcage, shoulders, legs, neck, and head remained, though some of these parts had been nibbled by small predators and rodents; birds had also pecked at the free meat. The animals that made these forays had removed much of the debris that the cat had piled over the remains when it was finished eating.

About thirty-five yards from the kill, which lay in the center of a cluster of rocks where small spruces and aspens grew, I found a fresh scratch. The mound, when uncovered, revealed scats that were still damp and tacky. This convinced me that the mountain lion had killed the deer sometime between midnight and dawn. Now I felt almost sure that the cat would return to the carcass during the coming night. So would I.

DARKNESS COMES EARLY AND QUICKLY DUR-
ing autumn in the western mountains, so
I returned to the lion kill at four in the
afternoon, carrying a flask of coffee, a con-
tainer of cold beans—previously cooked with
a few spices and a touch of garlic powder—
and my usual bag of oatmeal, raisins, and nuts. The coffee
and beans were for supper, the trail rations for snacking during
the vigil, which I expected would be long and somewhat
tiresome. I also had a canteen of water.

Scouting the area on arrival, I chose an observation point
sixty paces from the kill. From here I would have an unobstructed view. The place was upslope, within the perimeter
of an old slide that had thrust together a number of large
granite slabs in such a manner that they had created a shallow
depression in their midst, a little shelter in which I could sit
comfortably, facing the kill, my back propped against a flat
piece of stone.

In expectation of the chill that would arrive with sunset,
I had brought my down sleeping bag, and this, once I had
removed all the loose stones from the spot where I would
wait, I spread on the ground, opened wide so that one half
would be beneath me while the other half could be silently
flapped over legs and torso when I started to get cold. I had
put on two pairs of wool socks; over them I wore my winter
mukluks, which had been deliberately purchased two sizes
too large so as to accommodate thick felt liners as well as the
extra hose. My pants were mackinaw-type, made of heavy,
dark gray wool that resisted dampness and offered maximum
warmth; I had donned a thick cotton shirt and a heavy black
sweater and had brought along my down-filled parka (also
dark in color) and my winter hat, to be used later if needed.

When the food and drink, extra clothing, and my five-cell
flashlight were placed in readiness near the spread sleeping

bag, I walked down to the marsh and scooped up a handful of black mud, which I applied to face and forehead, first smearing a light coating over the exposed skin and then tracing abstract patterns with more mud; I didn't worry about my hands because I had a pair of dark gloves. Altogether, I must have looked rather like a commando raider, about to launch a night attack.

There now remained one last thing to do: I had brought the shirt and undershorts I had worn the previous day, and I now positioned them on the ground in the neighborhood of the kill, the shirt near the marsh, the underwear about fifteen yards south and west of the partly eaten carcass. If the puma came, its sensitive nose would lead it to my soiled garments; it would investigate, approaching slowly and cautiously until it was close enough to determine that neither object posed a threat. Then, when it went to feed, my body odors would be familiar, logged within the cat's brain as harmless, so if it picked up my scent later, it would probably ignore it, believing that it was again detecting the clothing. From this night onward, I planned to leave a number of such decoys in the territory of the mountain lion to instill confidence in the animal's mind. I wanted that lion to learn, and as quickly as possible, that the human in its territory could be trusted, that I was a harmless organism that would share the environment in peace.

By now the sun had set behind the peaks and the valley was beginning to fill with shadows. It was noticeably cooler; a light but persistent wind was traveling the gulch, coming from the south. The wind was perfect for my needs because it would fan over the old kill as well as over the garments and, more important, over me. If, as I suspected that it might, the lion returned from the north, the direction in which its departing tracks led, it would scent me and the clothing, but because my body odor would be stronger on shirt and underpants, it would investigate these first and might well fail to detect me once it started eating, for the wind would then be in my favor.

Before taking up my station among the rocks, I walked south for a quarter of a mile, settled myself on a boulder beside French Creek, and smoked my pipe, the last that I could have until the vigil ended. At 6:00 P.M., with darkness already smothering the valley floor, I was settled in my place of vantage. The sky remained clear and already the early stars were marching across the bluish void. The moon was now approaching the last quarter and was due to rise at about midnight. I lay relaxed, hoping that the lion would not arrive until after the moon came to give me some light.

The puma is a mysterious cat, an animal of the night and of the shadows, silent and cautious as a rule, but exceptionally noisy when moved to utter its fearful cries of love or rage. Walking on well-padded paws, it makes but the merest whisper of sound as it travels through its range, lithe and graceful and perhaps more alert than any other North American predator.

The lion prefers to hunt between dusk and dawn, and usually spends the daylight hours resting in concealment. But if it has not hunted successfully, if it is motivated by the breeding urge, or if it is disturbed, it will readily travel during full light. Sometimes, made restless by the heat of the summer or by pestering flies, it will cover considerable distances during the day, pausing to stalk game, or even stopping to investigate some tantalizing scent. Nevertheless, this wraith of the wilderness is always difficult to detect and for this reason it is one of the least studied animals on this continent.

Led by trained dogs, man can often locate a puma, but then the cat will run, sometimes seeking safety in a tree, on other occasions moving right out of its territory. Or it may turn at bay and attack the hounds and be shot by the pursuing hunters. To understand the true nature and character of this lovely savage, it is necessary to stalk it and observe it within the wild places that it favors. This is a monumental task that requires much time, considerable patience, a thorough knowledge of the forest, quick eyes, a lot of experience in the art of tracking, and a measure of luck.

Time can be found; patience and tracking experience can be learned, just as one can get to know the ways of the wilderness. Even luck can sometimes be influenced. But there is a final requirement imposed on any man who is seriously interested in studying this animal, one that is probably the most difficult to fulfill: the job demands *total* personal commitment. If one is fastidious or squeamish, if one loves comfort unduly, or if one is not prepared to follow wherever this elusive quarry leads, failure becomes inevitable.

The would-be puma observer must be truly interested in the cat; he must feel sympathy for it and relate to it fearlessly. There must be no gun in hand when stalking through the forest or during a lonely vigil in the dead of the night, for a weapon alters the psychology. A man becomes arrogant when he is armed; consciously or otherwise, he sees himself as the master of Creation. And yet, in contradiction, the fact that he feels the need to carry a firearm signifies that he is afraid. The rifle becomes his crutch; with it, he stalks like a killer, forgetting that he should be moving as an observer, with *all* his faculties focused on the quiet pursuit of knowledge. It is impossible to be at peace within oneself while carrying an implement of destruction. And it is imperative for one to achieve full inner tranquillity if one expects to understand the nature of wild animals.

I had studied mountain lions before, in other places, during a series of expeditions that would total three and a half years, I estimate that 50 percent of my time was spent studying the puma's tracks, kills, scratches and trails; 20 percent was used traveling and preparing myself on site; and only 30 percent of the surveillance hours gave me an opportunity to study the animal itself.

Most of my field work was carried out in British Columbia, but I also studied the puma in Florida and Mexico. In every instance, once I was in cat country, I worked alone and on foot, except in those areas where I needed a canoe to travel over lakes or rivers, or, as I did before starting my study in the Selkirk Mountains, when I used an aircraft for brief surveys. While some observers use radio telemetry as a means of studying the puma, I have always rejected this method. Before such technical equipment can be used, the lion (or other animal) must first be captured or darted with tranquilizers, weighed and otherwise examined while it is unconscious, and then fitted with a special collar to which a transmitting device is attached. Afterward, the observer follows the lion, by foot or in the air, picking up the radio signals with a portable receiver. The data obtained is often questionable: frequently the animal manages to rid itself of the disturbing collar; sometimes the yoke may interfere with its hunting, or get snagged on a tree branch, causing serious hardship—even death. But, more important, those who engage in radio telemetry are conducting *impersonal* electronic studies; scientific data is logged and noted by these observers, but they are unlikely to be able to fully relate to the subject of their study, to appreciate personally the animal's real nature, the hardships imposed upon it by the environment, its wants, and its needs. More likely, these men and women of science see the cat as a specimen rather than as a fellow being with whom they may share the wilderness world.

Sitting within my granite shelter that night, I was occupied by these thoughts for a time, but toward 10:00 P.M. the noise made by an animal traveling through the forest attracted my attention. It was not the puma; I knew this by the sound, yet I could not readily guess at the intruder's identity. It was approaching from the south, angling toward the kill at a slow rate; this suggested that a lone animal was either traveling with extreme caution, or else was unable to move more rapidly. Since my eyes were useless, I allowed my ears to track the stranger's progress. Some moments passed, then I heard a call: a low, moaning sound that was instantly familiar. A porcupine was coming my way. The identity of the animal was further confirmed by its rate of travel, for these docile creatures never move quickly. As the porcupine neared the partly eaten carcass, I reached for the flashlight; when I switched it on, the bristly animal was already nibbling at the remains of the dead deer.

I had seen porcupines gnawing on cast-off antlers, but I had not hitherto observed them eating meat, although I was not surprised that they would do so. Many rodents indulge themselves in this way when the opportunity arises, including such animals as hares, mice, and even beavers.

The porcupine chewed audibly, occasionally pausing to utter its rather mournful calls, a series of low, grumbling sounds typical of the species. It is as though the animals enjoy talking to themselves, or perhaps they wish to break the monotony of their solitary existence—a little puzzle that I have not solved and that continues to intrigue me. In any event, the porcupine did not stay long at the carcass, waddling off into the forest six minutes after it arrived. But shrews replaced it at the kill, their presence advertised by outbursts of shrill little cries uttered when one or the other of these carnivores encroached on its fellows and was attacked or chased away. I couldn't see the tiny, ferocious hunters, but I presumed them to be masked shrews (*Sorex cinereus cinereus*), for this subspecies occupies the greatest range in British Columbia.

Time passes slowly when one stands vigil in the dark of night. To speed it along, I usually attune myself to the sounds and scents of the environment, seeking to build a mind picture of the unseen wilderness. But that night my thoughts continued to dwell on the puma as I reviewed my findings in the valley. Then, when the moon approached the rise, the pale green, ethereal glow began to fill the forest, transforming it slowly into a place of vague shadows that teased the vision. When my eyes detected some formless etching that was but a tone or two lighter than the predominating darkness, it subtly altered shape or disappeared. At other times, as I tried to recognize an outline, the moon-wraith seemed to acquire physical characteristics, as though it were alive. I knew that I was experiencing optical illusions, just as I was aware that even in daylight a fixed stare can invest with life such things as an old stump or a cluster of moving leaves if they are seen from afar, but I could not stop myself from concentrating on each new and evanescent shadow. At least half a dozen times I imagined that I was looking at the crouching shape of a mountain lion. I even endowed with feline stealth a number of night sounds that were in fact made by mice or by tree branches rubbing gently against one another. Such fancies managed to spook me.

I began to think about the puma's reputation as a mankiller. This revived the memory of a trip that I once made to Pennsylvania in order to visit the grave of a man killed by a puma in 1751. His name was Phillip Tanner, and he is buried in the old cemetery in Lewisville, Chester County, near the border of Maryland.

I had waited years to see the tombstone that marks the late Mr. Tanner's resting place, becoming intrigued because this was the first documented attack on a European by a puma. So, when one day I found myself in Washington, D.C., on my way back to Canada, I took a detour through Lewisville.

There I learned that the puma's victim was killed at a place called Betty's Patch, which is conveniently located about half a mile from the cemetery.

Although details of the fatal encounter between man and mountain lion are historically sketchy, the victim's grave marker makes up for this lack. It is a fairly plain memorial with straight sides, a scalloped top, and a hand-carved inscription on the face that says: "Here lyeth ye body of Phillip Tanner who Departed this life May 6 1751 Aged 58 years." Above the relatively crude lettering, not quite in the center of the stone, the mason also carved a catlike figure shown with claws spread, extra long tail carried well over the back, the body crouched in an attitude suggesting attack; the carving is rudimentary, but the likeness to a lion is unmistakable. This is an interesting example of early tombstone art, when masons often tried to illustrate the cause of death if this was considered dramatic. By contrast, beside Phillip's grave, another Tanner was buried in 1761, but this headstone carries only a carved inscription that identifies the deceased and gives the date of birth.

To the best of my knowledge, this fatality represents the earliest authenticated puma attack in North America. Here and there through the literature I found old reports of violent encounters between man and lion, some laconic, others fancifully loquacious, but none offering sufficient proof. There are, however, contemporary records that prove that the mountain lion will attack, kill, and eat humans, though the cat does so only on rare occasions. That the animal can easily overcome a full-grown man is beyond question. It is enormously strong, agile, and fast during a short charge. Each large front paw is armed with five razor-sharp claws; the hind feet carry four each that are not so sharp, but are used to rake and disembowel the prey while the great fangs and the ten front claws hold the victim in a death grip.

Theodore Roosevelt, who hunted mountain lions many times, likened this cat to the panther. He believed it to be "quite as well able to attack man; yet instances of it having done so are exceedingly rare. But it is foolish to deny that such attacks on human beings never occur." And he stressed: "[W]e must never lose sight of the individual variation in character and conduct among wild beasts."*

Those words were written before I was born. I was not to read them until I had amassed wide experience with wild animals in North America and elsewhere and had myself independently arrived at more or less the same conclusions. Every animal is an individual. Although it is linked to members of the same species by an undetermined number of genetic traits and thus shares certain basic and common characteristics, its life is largely governed by its own precepts. As a result, it is never safe to generalize about the behavioral patterns of any animal, for, even among members of the same family, individual differences are clearly evident. From mouse to man, we all enter the world equipped with basic genes that will allow us to live according to the dictates of our biological needs, but how we live, how we use our inherited endowments, depends largely on our experience, on our personal abilities to learn, on our own highly individual personalities, and particularly upon the influences (negative or positive) exerted upon us by our parents and siblings, to name but some of the factors that blend to build a personality.

Applying these concepts to a predator as powerful and skillful as the puma, one may wonder why this meat-eater does not regularly stalk and kill man. The reason, I believe, lies in the taste buds.

For several years I became involved in a study of attacks

*Theodore Roosevelt, "With the Cougar Hounds,"part 1, *Scribner's Magazine* 30 (4) October 1901, pp. 432–33.

made on man by a variety of large predators, records of which are to be found in every country in the world. Many of these accounts are unfortunately too brief, merely recording the killing and the animal that launched the attack, but there are more than enough detailed reports that, when examined carefully and compared with one another, demonstrate that the greatest number of attacks by far have been prompted by three circumstances: The animal was starving, was provoked in some way, or was ill or injured. Only a minority of documented killings suggested that an individual predator attacked for other reasons; but in these cases, either the animal in question was not discovered, or if it was, it was not thoroughly examined—that is to say, it was not autopsied in an effort to seek evidence of illness, injury, or deformity.

I conducted research over a period of years, starting while I was still living in Europe and continuing in Africa. From India, a country I have never visited, I obtained reports from individuals and from the literature. In the Americas, I studied old and contemporary records and talked with officials and individuals who had personal knowledge of some of the attacks. In more recent times, I compiled my own statistics. Searching the records, I found that 609 cases of *documented* attacks on humans by large carnivores had occurred between 1751 and 1958. (I am sure there were a lot more that I missed.) Five hundred twenty-seven of these resulted in death. The animals involved were African lions, panthers, leopards, tigers, European wolves, pumas, and grizzly and black bears, in addition to only one wolf attack on a railway worker in Ontario.

In thirty-two of these cases, including that of the Ontario wolf, the animals were found to be suffering from rabies. To this list I can now add two more attacks of which I have personal experience. One was made on me by a grizzly bear in northern British Columbia in 1971. Forced to kill the animal in self-defense, I discovered that it had been wounded about a week before it blundered into my encampment—using an aircraft, hunters had fired on the bear, spraying its back and haunches with buckshot. The second attack was made by a black bear against an Italian tourist on the Alaska Highway south of Watson Lake, Yukon Territory. In this case, the inexperienced man was cycling north, carrying food in his pack. The bear was standing beside the road and the tourist stopped to photograph it; when the bear smelled the food and advanced, the man panicked and threw his bicycle at it. The bear then attacked. Fortunately a truck came along and the driver had a rifle. He shot the bear and took the tourist to a hospital, where he received one hundred stitches.

Before these last incidents, I had studied the taste of raw meat. Starting with beef, pork, and chicken, I sampled uncooked morsels from such diverse creatures as muskrat, beaver, deer, moose, bear, wolf, coyote, mink, and many others, securing my samples from trappers and hunters. The task was distasteful, but I was able to determine that each kind of meat has its own distinctive flavor. Of course, I have not sampled human flesh, unless I can count the times that I have sucked my own bloodied fingers! Nevertheless, it is not difficult to determine that all predators have what may be termed favorite prey animals that they hunt by choice when these are available. Should its number-one food become scarce, the carnivore will turn to the second or third choice, but if faced by actual famine, it will take whatever it can find, including carrion. Under normal circumstances, man is not a preferred food, but if a land carnivore is driven by need to attack a human, it may well become a confirmed man-eater, probably because our species is so very easy to stalk and kill. This does not mean that the large predators dislike the flavor of human flesh. Indeed, some of them may thoroughly enjoy it! But man is not regarded as normal prey.

The young carnivore, like the young human, acquires its food habits in the nursery; it must eat what is brought to it. Later, when old enough to accompany the adults on the hunt, it learns to identify preferred species by scent, sight, sound, habitat, and behavior, relating these factors to the taste of the meat. In this way the animal becomes expert in the ways of a relatively few selected species, seeking them actively and always able to detect them by the signals they emit, be these vocal, odoriferous, physical, or habitual. As the predator matures, it learns to recognize other acceptable prey, widening its dietary choices in the face of need, but usually seeking those animals to which it has become habituated. The signals that man unconsciously emits as he travels through forest or jungle do not normally trigger appetitive urges in the mind of a carnivore. Man is himself a predator; he does not move or behave like a prey animal, and this, coupled with his alien odor, is more likely to inhibit rather than tempt a meat-eater. Nevertheless, no powerful animal of any species should be taken for granted, for even prey animals, such as moose, elk, or bison, are capable of killing a human if they are provoked.

Sitting within my rock shelter reviewing these things, I succeeded in making myself fearful to the point where I was tempted to abandon the vigil; but then I realized that if I quit this night I would continue to feel afraid and might then just as well pack up and leave. I became stern with myself. I was no tyro, having spent a great many nights waiting for, and watching, wild animals. Never before had I managed to think myself into a deep funk. Chimeras induced by autosuggestion, I knew, are far more terrifying than reality, so I now engaged in my own personal ceremony of exorcism.

I turned my thoughts away from the puma and forced my memory backward, to a day during last summer when I had deliberately dived into the ocean while a 23-foot killer whale was but yards away from my boat. Klem, as I christened the bull, had arrived one morning in the bay where the *Stella Maris* was lying at anchor, off an inlet in northern British Columbia waters. For days the whale kept returning and I fed him tidbits saved from my fishing trips, often dropping the food into his cavernous mouth when he "stood" on his tail beside the bow rail. Then, one morning, I felt it was time to fully trust the great mammal, so I put on my scuba gear and went over the side, at first feeling apprehensive, but then reminding myself that the only way to relate to an animal is to meet it on its own terms. Klem returned my trust and, in doing so, reminded me of my dog Yukon, who was utterly savage when he came to me, but responded when I banished my fear of him and exposed myself to his fangs.

Such mental images restored my equanimity. By the time the quarter moon appeared, I was totally at ease. That night I relaxed so well that I fell asleep some time after 1:00 A.M.

The first European to mention the puma was Columbus, who saw the animal during his fourth voyage to the New World in 1502, during his explorations of the coastlines of Honduras and Nicaragua. He called it "lion," as did later Spanish explorers. One of them, Bernal Diaz del Castillo, went with Cortes to Mexico and relates that in 1519, when the Spaniards entered Montezuma's stronghold, they found a "zoo" in which the ill-fated Aztec ruler kept many of the native animals.

Describing the city of Mexico as it was then, Bernal Diaz wrote: "Let us go on to another large house where they [the Aztecs] kept many idols whom they called their fierce gods and with them all sorts of beasts of prey, tigers and two kinds of lions. . . . They were fed deer, chickens, little dogs, and other animals which they hunt and also on the bodies of the Indians they sacrifice, so I was told." Later the old *conquistador* said that many of his comrades were fed to "the lions."*

It is not clear what this chronicler meant when he referred to "two kinds of lions"; when he talked about "tigers," he meant the jaguar, which to this day is known as *el tigre* in Central and South America, but he was obviously under the impression that he had seen more than one kind of puma. Perhaps Montezuma had captured one of the rare albino cats, or a black puma, equally rare.

No other carnivore in the Americas is known by so many different European and Indian names; not counting its Linnean or scientific title, this animal is known by no less than seven English, two Spanish, and two Portuguese names.

In addition to being called mountain lion, puma, cougar, and panther in English, the animal is also known as the Mexican lion, catamount, deer tiger, and painter (a corruption of panther). Brazilians refer to it as *onca vermelha*, or *leão*. In Spanish it is called *león* and *leopardo*, but it is also still known by many Indian names. The animal is *chimblea* in Baja, California: It is still known by its Aztec name, *mitzli*, in some parts of Mexico; the Apaches christened it *yutin*, the Ojibways *mischipichin*, the Sioux *ig-mu-tan-ka*; the Mandans named it *schunta-haschla*, the Omahas *ingonga-sinda*, and the Osages *ingronga*. In Chile, the cat is *pagi* or *paghi*, and some Spanish chroniclers called it *tigre rojo*, or "red tiger."

Two South American biologists, Angel Cabrera and José Yeppes, point out that the name *puma* is widely used and

Historia Verdadera de la Conquista de la Nueva España (Madrid: Editorial Porrua, 1960).

has even been adopted in some countries to describe the genus. They also say that "[the name] is Quichua; in Araucana the same animal is called *paghi* or *trapial*; in Puelche, *haina*; in Guarani, *quasuara* or *yagua-pihta*; and in Tupi, *sussuarana*. . . ."* The name *cuguar* (cougar), which appears in some natural-history works, is not a word in any native American language, but is a book term coined by Buffon,† who was very fond of disfiguring such names of animals as he found difficult. He thus transformed the term *cuguacuarana*, a capricious transformation in turn from *sussuarana* (Tupi Indians), copied by the early traveler Marcgrave.‡

But the confusion surrounding the cat's nonscientific names pales by comparison with the complexities of the animal's scientific nomenclature, for no less than thirty subspecies of *Felis concolor* have been described and catalogued over a vast north-south range that stretches from northern British Columbia to 55° south latitude at the Strait of Magellan, the extreme edge of the South American mainland in Chile, while its east-west habitat encompasses both the Atlantic and Pacific seaboards.

Thus, in northern British Columbia, northeastern Washington, northern Idaho, Montana, and northwestern North Dakota, we find *Felis concolor missoulensis*, the cat that I waited for that night, which was "discovered" in the area of Sleeman Creek, Missoula County, Montana, and described by Goldman§ in 1943; while in southern Chile and in south-

---

*Angel Cabrera and José Yeppes, *Historia Natural Ediar. Mamíferos Sud-Americanos: Vida, Costumbres y Descripción* (Buenos Aires: Compañía Argentina de Editores, 1940).

†George Louis Leclerc de Buffon, *Histoire Naturelle* (Paris: 1761).

‡G. Marcgrave, *Historiae Rerum Naturalium Brasiliae* (1648).

§Edward Alphonso Goldman, "Two New Races of the Puma," *Journal of Mammalogy* 24, June 8, 1943.

ern Argentina lives *Felis concolor patagonica,* described by Merriam. *

Scientists, being purists by necessity, keep probing, dissecting, analyzing, and measuring; let there be but one tiny bone in animals in one locality that differs however marginally from the same bone found in the same species in other ranges, and a new subspecies is announced.

This is very much the case with *Felis concolor.* Bone structure is the only more or less reliable means of telling one puma from another, but even this method has drawbacks. *Felis concolor* interbreeds freely. A lovelorn tomcat whistling his way through the forest isn't too likely to worry about species niceties if he is answered by the amorous yawl of a lady puma; and the young of a union between, say, *F. c. missoulensis* and *F. c. oregonensis,* whose ranges overlap, may inherit the anatomical characteristics of one parent, a mixture from both, or a compromise between the two. If the last is the case, then a "new" subspecies is destined to roam the forests while waiting for posthumous fame to be bestowed upon it by some probing biologist.

What is the color of a puma? One may almost take one's pick: red, brown, tawny, yellow, gray, slate-gray, white, or black. But none of these shades will do in themselves, for, the last excepted, there is a wide variation in hues and tones, many of which almost defy description. Theodore Roosevelt and some of his friends killed fourteen pumas in 1901 (the year he became President) during a hunting trip in Rio Blanco County, in northwestern Colorado. He reported that the animals showed "the widest variation" in color. "Some," he

*Clinton Hart Merriam, "Preliminary Revision of the Pumas (Felis Concolor˙ Group)," *North American Fauna* 16, Bureau Biological Survey, U.S. Department of Agriculture, 1901.

said, "were as slaty gray as deer when in the so-called 'blue,' others rufous, almost as bright as deer in the 'red.'"* The famous big-game hunter added that the color of these animals did not appear to have anything to do with sex, locality, season, or age.

Size can be equally deceiving. In 1928, Seton described a record-sized puma. It had been killed in 1917 by J. R. Patterson, a government hunter, in the area of Hillsdale, Arizona, and weighed 276 pounds *after* the intestines had been removed. "He measured 8 feet 7³/₄ inches. . . ."† Alive, this tom must have weighed more than 280 pounds.

Roosevelt, who was a painstaking and dedicated hunter-naturalist, has left us a record of the 14 pumas taken in the United States in 1901. Two of the animals were young females, weighing 47 and 51 pounds, respectively. Nine adult females measured between 6 and 7 feet and their weights varied from 80 to 133 pounds. Three males killed were larger and heavier. One was 7 feet 6 inches long and weighed 160 pounds, another measured 7 feet 8 inches and weighed 164 pounds; the third was a record puma, 8 feet long and 227 pounds.

These and other statistics suggest that adult pumas do not often weigh more than 200 pounds or less than 70, and that if there is such a thing as an average weight, this probably ranges between 100 pounds for females and 150 for males, which are usually between 20 and 50 pounds heavier and 12 to 24 inches longer than their mates, but there are so many variables that generalizations mean little. Habitat, availability and kind of prey, time of year, weather conditions, age,

---

*"With the Cougar Hounds," part 1, *Scribner's Magazine* 30 (4) October 1901, p. 435.

†Ernest Thompson Seton, *Lives of Game Animals*, vol. 1, part 1 (Garden City: Doubleday, Doran & Company, 1929).

injuries received while making a kill, number of young in a litter, parasitism, disease, and inheritance can all influence the size, weight, and color of individual animals.

The puma is unquestionably adaptable. It is equally at home in the high mountains, the torrid jungles, the desert, or at the very edge of the sea; it is so adaptable, in fact, that it is only intolerant of the open plains or of habitats drastically altered by man. Even so, it is clever enough to live unnoticed in the very shadow of humanity, provided that it is not tempted to attack domestic stock and thus give itself away.

Unfortunately the big cat usually cannot resist helping itself to cattle, sheep, pigs, horses, even barnyard chickens, and it will not disdain a farm dog. *Felis concolor* all too often stumbles over domestic stock and discovers that these strange creatures are easy to kill and good to eat.

Hunting its natural prey under natural conditions, the cat must work hard before it can go to sleep on a full belly, and it is by no means unusual for it to go hungry for days at a time.

In the wild, if a deer is pulled down, none of its companions are going to hang around to give a predator a second chance. But domestic stock are slow and invite slaughter. Sad to say, *Felis concolor* must always be stopped when it begins to help itself to domestic stock, but nothing can justify the persecution of this animal within those wild parts of its range that are far from the domain of man.

Laws for the cat's protection have been enacted in most parts of North America, and one might hope that these will ensure the animal's continued survival, but a law is good only if it can be enforced, and neither the United States nor Canada has a sufficient number of rangers or game wardens to effectively police the wilderness. This lack of personnel allows brutal, unscrupulous people to poach pumas—and many other animals—by every means at their disposal, including the use of snowmobiles, air boats, aircraft, and ground-to-air radios, against which the animals stand no chance at all.

Since the beginning of the present century, the eastern puma, *F. c. couguar*, and the Florida subspecies, *F. c. coryi*, have been considered extinct or close to extinction, thereby earning for the animal a place in *The Red Book* of the International Union for Conservation of Nature and Natural Resources.

Of the eastern puma, *The Red Book* says: "Formerly ranging from the eastern United States to provincial Canada west to the edge of the plains and Alberta, this race has been extinct in the United States since the end of the last century. It was almost eliminated in Canada as well, but although still rare, it is now believed to be gradually increasing its range."[*]

Young and Goldman[†] confirm that the eastern puma is regarded as being extinct east of the Mississippi River, but Cahalane[‡] wrote in 1944 that the U.S. National Park Service had received reports of puma sightings in Virginia's Shenandoah National Park; more recently (in 1979) I talked to a U.S. Fish and Wildlife Service biologist who told me that he had received word that year of more sightings in Shenandoah. These hopeful reports coincide with recent sightings in eastern Canada.

On two occasions (1954 and 1963) I found puma tracks in Ontario. In both cases the marks were so clear that they

[*]James Fisher et al., *The Red Book: Wildlife in Danger* (London: William Collins Sons & Co. Ltd., 1970).

[†]Stanley Young and Edward A. Goldman, *The Puma, Mysterious American Cat* (New York: Dover Publications, Inc.; originally published by American Wildlife Institute in 1946).

[‡]Victor H. Cahalane,Letter to U.S. National Museum (Washington, D.C.: 1944).

could not be mistaken for the sign left by any other member of the cat family. In the first instance, the animal had crossed the old Trans-Canada Highway in the Hearst-Longlac region during the night of December 2, 1954. The temperature was below zero, the snow two feet deep, hard and compacted except for some two inches of fresh fall that had accumulated the previous evening.

I had left Hearst at five in the morning, bound for the town of Longlac, 130 miles away. At about seven-thirty I stopped the car in order to stretch my legs and brew some coffee, for in those days there were no restaurants or gas stations open during winter along the entire stretch of roadway that separated the town I had left from the one for which I was bound. There were plenty of dead spruces on either side of the highway, so it didn't take long to get a fire going, fill the coffeepot with snow, and hang it over the flames from a tripod of green sticks.

While the water was heating, I walked, staying on the southern edge of the forest and trying to keep myself from breaking through the crust. Although it was not yet full daylight, visibility was good because of the clear air and the reflective quality of the snow. Since this was the first time I had set foot inside a boreal forest, I soon became fascinated by my surroundings.

I hadn't gone very far when I saw the tracks. The animal had come out of the trees on the north side of the road, jumped the small ditch, and entered the south side of the forest. Inside the spruces, the lion had sunk belly-deep where it landed, then it had scrambled up and walked on top of the crust. In several places it had dragged its long tail in the snow. On each side of the highway, the prints left by its great paws were clear and sharp.

I was not then aware that the puma was supposed to be extinct in Ontario, so I was not surprised by the tracks, but even though I was a greenhorn, I had enough sense to realize that I was being treated to an unusual sight, and I measured the footprints and sketched them to size. The marks of the front feet were $3^3/_4$ inches wide and $3^1/_4$ long, those of the back feet were $3^3/_8$ inches wide and 3 inches long, and the width of the trail, or straddle of the animal, measured 11 inches. The straddle is the distance between the animal's paws measured from the inside edge of each pair of pads; it marks the *width* between the feet at each stride. Obviously this measurement varies in accordance with the size of the animal and the condition of the trail, which may sometimes cause the cat to bring its feet closer together, but on good level ground, during normal, leisurely walking, the straddle remains fairly constant; in the puma, it runs between 8 and 14 inches. By contrast, the straddle of the lynx, whose big pads can be confused with those of *Felis concolor*, measures between 7 and 9 inches and averages 7 inches.

I sighted the second set of tracks in mid-July on the east bank of the Missinaibi River, near its junction with the Fire River, an area approximately two hundred miles north of Lake Huron's North Channel. The rough bearing taken at the time gives the position at 49° 00′ N by 83° 15′ W. In this instance, the tracks were in fine, damp sand, an ideal matrix that offered some sharp imprints additionally supported by a scratch.

Uncovering the scats and spreading them with the aid of two sticks, I noted that they were full of porcupine quills and fur, numerous chips of bone, and other substances unidentifiable with the naked eye.

The tracks were slightly bigger than the set I had seen in 1954, measuring 4 by $3^{11}/_{16}$ inches (front) and $3^9/_{16}$ inches by $3^1/_8$ (rear).

Sketchy though the evidence may be, it suggests that the eastern puma is making a modest return to some of its former ranges. The increase corresponds with an even greater rise in the population of deer, which have become abundant in the hinterlands of eastern North America during the last twenty years because of stepped-up logging operations. (When the forests are thinned, the shrubs and seedling trees on which the deer thrive in winter quickly replace the mature trees, and grow in profusion.)

Unless one is stubbornly unbelieving, one must accept that the puma has been recently sighted in areas where it was thought to be extinct; one must also accept the finding of scats and tracks by an increasing number of trained investigators. Having twice seen puma tracks in the northeast, I certainly credit the evidence. But I do not believe that the lion was ever extinct in these ranges.

It is my view that the animal dwindled steadily over the years until it reached a point of near-extinction and was thereafter extremely difficult to detect. At this stage, the big cat probably retreated into the deep forests and there continued to survive without attracting attention. In more recent times, the abundance of deer naturally led to an increase in the population of pumas, while a spectacular rise in the numbers of nature-lovers has led to an increase in sightings.

I DON'T KNOW WHETHER IT WAS THE NOISE or the chill of a mountain dawn that awakened me as the first blush of sunshine turned the valley into a place of pink, ethereal light. The fact is that my eyes opened and noted the new day while at the same time my ears detected a persistent scratching sound coming from nearby. For some moments I lay unmoving, absorbing the news that consciousness was giving me, but not quite able to interpret the messages. Presently—and I don't know how long it took me to collect all my faculties—I understood that I had slept through the night and that the sound originated from a place inches from my head.

The noise stopped, then became magnified suddenly and was accompanied by a shrill squeak. I started. I was hardly able to collect my wits before a pack rat erupted from inside my bag of trail rations and streaked away, scolding angrily, the swift pattering of its feet seeming to match the tempo of my accelerated heartbeat. I had left the canvas sack beside my sleeping bag, its neck still closed by the purse-string, but one of its sides showed a neat, almost perfectly round hole about two inches wide. A piece of the material that exactly matched the new opening lay a few inches away, its edges roughened by the rat's sharp teeth.

By now I was fully awake, struggling between a desire to laugh at the cheek of the little raider and an urge to become angry with myself for having slept right through the night. When I sat up, anger won, for I saw at once that the puma had returned and had eaten an amount of meat from the kill. It had even dragged the carcass some distance away from me!

The morning was cold, but the chill didn't bother me as I jumped up and walked down to the remains, my disappointment so intense that it dominated all other feelings and emotions. "*Damn, damn, damn!*" I swore aloud, really angry now as I stopped beside what was left of the deer. The cat had consumed a large amount of the remains; it had even cracked the skull and licked the brains out of a gaping hole left when the head was severed. Fang punctures through the orbital arch and the frontal ridge measured $5/16$ of an inch in diameter, suggesting that the cat had bitten through until its canines penetrated up to the gumline.

My anger disappeared as I busied myself examining the carcass and later searching the ground for tracks. Many of the rocks in the area were in places covered by thin soil that was topped by mosses and small plants; but in other locations the granite was concealed by a layer of fine sand, rock dust deposited by winds or produced by erosion. The sand preserved the cat's prints beautifully; the mosses, though they showed the rough outlines of the pads, had to be given close scrutiny before I could determine the direction of travel.

Walking and crawling carefully so as not to disturb the signs, I spent about twenty minutes checking the ground in the immediate vicinity of the kill, slowly widening my area of search. At first I backtracked the cougar, confirming that it had, indeed, approached from the north. Later I was pleased to find that the cat had checked the garments I had left as decoys. The shirt had been moved about fifteen feet from where I had left it, and the right sleeve had been torn in several places. The undershorts remained in their original location, but beside them, on the earth that I had myself bared the previous evening when I removed the moss from a rough circle encompassing the garment, were several large tracks left by the puma's front paws.

The lion had come down the valley, traveling on the east side of French Creek, but above the marshland; it had gone beyond the kill to first check the undershorts and then the shirt, and had afterward walked directly to the carcass. I could not, of course, determine whether it had moved this immediately, or at a later time.

It took me more than an hour to uncover the available evidence and form my conclusions. By then I had come full circle and was again standing beside the remains of the deer. Now I began to search for signs that might tell me the direction in which the cougar had gone after its meal. This was difficult because of the many tracks left by the cougar and other animals.

I went down on all fours and used the large magnifying glass I carry for such purposes (a behavior characteristic that once prompted a colleague to refer to me as a Sherlockologist), but even with this, it took me some fifteen minutes to determine that the cat had aimed itself at the place where I had been sleeping so peacefully!

Inch by inch—literally—I scanned the ground and worked my way in the wake of the pugmarks, sometimes losing the trail among loose stones and boulders, occasionally finding good tracks, more often than not having to trace the indistinct outlines with a sharp twig, noting the bruising of the moss and grass, the nearly indiscernible indentations made in the vegetation by the pressure of the lion's pads. By the end of the second hour, I had covered 217 feet (by later measurement) along a curving course that led from the kill toward my resting place, angling slightly south after forty-five feet, then continuing toward the east, running parallel with my position until, climbing a rise, the tracks stopped at the edge of a nine-foot drop *immediately above* the place where I had been fast asleep.

I would like to be able to report that I kept my cool and that the hairs on the nape of my neck didn't rise stiffly. I did not feel conscious fear, of that I am quite positive; yet I found myself gripped by excitement, my heart accelerating, those familiar "butterflies" corresponding to the clenching of my stomach muscles. In my mind, I visualized the lion arriving like a ghost, guided to me unerringly by its marvelous sense of smell, and stopping but a short leap away to inspect me in the moonlight.

Suddenly I became elated. The puma had found my garments, got my scent. It had pursued it after taking its fill from the carcass, had found me, inspected me, and then gone on its way.

Aloud I said to myself, "It didn't attack!" But the words had hardly been articulated when a small inner voice seemed to whisper, "Yes, but maybe that was because it had already eaten well."

I put the doubt out of my mind, remaining elated. Without being able to explain it rationally, I was sure that the big cat would not have attacked even if it had been hungry. And I felt that its visit was an omen, that this "ghost walker" and I were going to become much better acquainted.

It occurred to me that "ghost walker" was an apt description of the animal. I was not yet sure that the lion was a male, although the size of its tracks suggested that it was, but regardless of its sex, the title suited the puma so well that from then on I thought about it and referred to it in my notes as Ghost Walker, eventually shortening this to Ghost. As I was to learn, the name was more than apt!

Descending from the rock after examining the tracks, I returned to the deer carcass, seeking to determine the approximate quanitty of food that the lion had ingested, for I was most anxious to estimate the amount of meat that these cats consume annually. Up to that time, the available figures referred only to deer and ranged from a high of three hundred to a low of thirty-five animals. The maximum number was ridiculously inflated, I felt sure. The low estimate might be closer to the mark and tended to more or less agree with my own arithmetic, but I knew I would not be able to arrive at a relatively accurate estimate of the total meat consumed during a year of hunting until I concluded my study in the Selkirks.

I believed then that it was safe to assume that the puma's preferred prey animal is the deer, either the mule (*Odocoileus hermionus*), or the white-tail (*Odocoileus virginianus*), wherever these or any of their subspecies occupy the lion's natural range. But this does not mean that the puma rejects other prey, even when deer are plentiful within its habitat. Opportunism is a characteristic of all animals, for survival demands flexibility in matters of food and range. For this reason, a hunting cougar will take almost any animal that happens to present itself, if it is hungry enough; *its natural diet is not confined to just one species of prey.* In British Columbia, as in many other regions of North America, the puma also hunts moose, caribou, marmots, beavers, goats, sheep, porcupines, snowshoe hares, bear cubs, coyotes, mice, birds (especially grouse and ptarmigan), and even wolves, if given the opportunity to separate one wolf from the pack.

I had already established that the particular animal I was seeking to study had tried for a marmot and failed. Up to now I had evidence that it had killed two deer in the area of French Creek, as well as some other kind of animal. I came to this conclusion when I found traces of fine fur in the scats I had examined at the trail mouth during my reconnaissance trip to the valley of the shack. Now I was determined to list all the kills that I might find and to try and estimate from examination of the remains the amount of food ingested by the cat at one sitting.

I had not sought to attempt this with the first, older kill because not enough of the evidence remained. The second carcass would yield better information, but even this would be difficult to assess, for I really had no means of knowing how much of the deer had been consumed by the puma and how much by other animals scavenging at the remains.

Because of these imponderables, I began another scrutiny of the ground around the leftovers, this time looking for the tracks left by other animals. Almost immediately I found two distinct sets of coyote prints, one larger than the other; more than one porcupine had also come to feed while I was asleep. The minute tracks of shrews and mice were everywhere to be noted, and at least one weasel had stopped to enjoy a free meal.

All this information was duly recorded when, after three-quarters of an hour, I decided that I had discovered all the visible traces of the night prowlers. Then I began to examine the carcass, finding that from nose to tail it would have measured, alive, about 48 inches. This suggested that the animal had been a young doe, probably one of last spring's fawns, the live weight of which could not have exceeded 70 pounds. Allowing for dehydration after death, and for the weight of the bones, body fluids, and hair, I estimated that the amount of usable meat on the animal would probably have totaled between 40 and 45 pounds.

The puma had taken two meals from the carcass; I guessed that the other meat-eaters had also partaken of the remains. I estimated that the scavengers had eaten 30 percent of the total consumable meat, the lion 50 percent, and that 20 percent remained in the carcass. The final figures showed:

Total meat, 40 pounds; eaten by scavengers, 12 pounds; eaten by the lion, 20 pounds; remaining on bones, 8 pounds. These calculations suggested that the puma had eaten about 10 pounds of meat during each of two feeds, again a rough estimate.

Aided by my knowledge of the lion as well as by the more careful estimates made by other qualified observers, I found that my rough calculations compared favorably with the generally accepted view and that those who estimate that the puma takes thirty-five deer a year appear to be closer to the mark than those who believe that the cat kills between thirty-five and one hundred annually.

Averaging the weight of deer at 125 pounds live weight—so as to err on the side of caution—I next deducted 30 percent from this weight to allow for dehydration, bones, body fluids, hair, hoofs, and contents of stomach and intestines. This left a total of approximately 87.5 pounds of meat on each deer. Thus, fourteen deer could be expected to yield a minimum of 1,225 pounds of edible protein in one year. If a lion consumed 50 percent of that total (612.5 pounds) it would need, ideally, twenty-eight deer a year. These were very rough figures, admittedly, but they were as close as I could come at that moment.

However, I felt then (as I do now) that to seek to determine the exact number of any given kind of prey animal that a predator kills during the course of one year is akin to trying to paint a picture on a cloud. I preferred to try to discover the total amount of meat from all sources that a cougar *needs* to sustain itself in relatively good condition. Even this was a monumental task, for the only truly reliable figures I had came from zoos, where a lion does not have to earn its own living. Some zoos that I had seen in Europe consistently underfed their captives, while a few tended to overfeed them.

Just as man does not live by bread alone, a mountain lion does not survive solely on deer, or on any other kind of animal. A particularly fortunate cougar, living on a range where mule deer or white-tail deer have overpopulated themselves, could well kill fifty or even a hundred of these animals during the course of one year, forming the habit of eating its fill at each kill and leaving the remains for the scavengers. But such an ideal situation would not last very long, because wherever one species of prey animal reaches a population peak, the predators multiply right along with it and soon restore a balance between their own numbers and those of the prey species.

No, a mountain lion, efficient hunter though it is, nevertheless lives from feast to famine, as do most carnivorous animals, so I felt that the best I could hope for was to obtain a reasonably accurate estimate of animal food consumption. If, during the course of the study, I was also able to form an opinion as to the kind and quantity of animals that were killed by the lion, it might be possible to *guess* at the numbers of large prey species generally taken by these cats. Such information might then be useful to biologists who seek to establish the ideal prey/predator balance on a given range in order to introduce meaningful conservation policies.

On that October morning I did not dwell for long on such matters, employing myself instead in making notes of the evidence, in the hope that time might give me some answers.

Because the pack rat had stolen most of my dry rations and I had taken so long to examine the kill and its surroundings, I was ferociously hungry when I returned to my base a little before noon. Here I found Wisa, Ked, and Jak waiting for me impatiently, all three sitting on the roof of my shelter. We had lunch together, each bird first busying itself eating those trail rations that the rat had not filched and afterward taking bits of bannock bread from my fingers and flying away to hide the food in the forest.

When we finished the meal, I spent an hour thinking about my quest, for I now realized that I was not as well organized as I had thought. To begin with, I had so far failed to properly explore the region; as a result, I had not done a survey of the other major animals that lived in the environs of French Creek and the Goldstream River, especially deer, although

I had noted a number of their tracks within that part of the French Creek valley that I had explored. But I did not know whether deer were plentiful or scarce, nor did I know where to find them. Similarly, although I had seen a small group of mountain caribou while flying over the area, I had not yet found any of their tracks. In short, I now had to discipline myself to do a thorough survey of the two waterways and of the lowlands through which they coursed.

I had also neglected to look for suitable places on which to construct several temporary shelters for those occasions when I was too far away from my base to return to it after a night's vigil, or at those times when I ended my observations at sundown. And I had not cut and stacked a supply of firewood for the winter.

It was depressing to contemplate the variety of tasks that still remained, until I faced the fact that I had allowed myself to become so anxious to actually see the puma that I had neglected to put into effect the plans I had made while I was still in Revelstoke. It was as though I had set out to climb a mountain, but expected to reach the peak without having to toil up the side.

Having given myself a silent but healthy scolding, I felt better, ready to start at the beginning and to prepare myself systematically for an effective study of the mountain lion. As a first step, I spent the remainder of that day cutting, splitting, and stacking firewood, a good supply of which was available close at hand in the form of lodgepole pines between 10 and 14 inches thick. Many of these had fallen, while others still stood, the bark gone; they were as dry as kindling and all of them were within easy carrying distance. By nightfall I had a stack of fuel that measured 8 feet long, 5 feet high, and 4 feet deep; this amounted to 160 cubic feet of wood, or about

1$^1/_3$ cords.* I knew I would need more before the winter was over, but I could now easily keep up with the daily consumption.

The next morning I set out before the sun had topped the peaks and headed downtrail toward the mouth of French Creek, where I had left the canoe. I carried enough food for six days, my tent, sleeping bag, saw, and axe, and a roll of heavy-duty polyethylene that I intended to use in constructing a bivouac somewhere in the eastern reaches of the Goldstream River; this would be the first of the emergency shelters.

My plans now called for a thorough exploration of the Goldstream up to the place where it leaves the mountains and enters the valley, which is about sixteen miles east of the mouth of French Creek; then I proposed to survey the western part of the river as far as the beaver pond where I had seen and heard the great horned owl, a location five miles from the sandbar I had used as a landing place. When these studies were completed, I would turn my attention to the French Creek valley, exploring it thoroughly from the sandbar to its terminus, a total distance of nine miles. But the mileage had to be doubled, for I had to explore both banks of each waterway. This meant that I was going to examine at least forty-two miles of Goldstream country, and some eighteen miles of the French Creek area, a *minimum* of sixty miles encompassing about thirty square miles of territory. I elected to do the river survey first because it was still navigable and by using the canoe I could save a lot of time. Nevertheless, I expected to do a great deal of walking, for it is not possible to make a thorough study of land by

*One cord of wood contains 128 cubic feet and measures 4 feet by 4 feet by 8 feet. I had in the past earned my living as a logger, so that it was not difficult to cut such an amount in five hours when the logs were so readily available, even using a bow-saw.

merely examining the shorelines of the waterway that bisects it.

Going upstream slowly, I devoted almost all my attention to the north shore, only turning to look at the south bank when something moved within the trees or on the river's edge. I paddled for ten minutes, landed, and spent time walking on shore and exploring an arc that was probably a third of a mile in length and a quarter of a mile in depth; then, after noting landscape, tree species, plants, animals, or tracks sighted, I returned to the canoe to repeat the process, in this way working upriver in easy stages and obtaining a good idea of the land and the life it contained. Later, after freeze-up, I would return to do the same journey on foot, this time broadening the area of search up to the point where the land begins to rise and forms the lower slopes of the mountains.

At noon, having explored about three miles of valley, I stopped for lunch beside a small waterway. According to the map, this was Spencer Bill Creek—whoever Spencer Bill was. Here I made my first important live sighting when a female wolverine, with two cubs, trotted sinuously out of some deep undergrowth and led her two-thirds-grown children to drink Spencer Bill's water. I sat still, only about twenty feet away from the animals—a ringside seat!

The wolverines knew I was there. The mother's small, rather rounded ears were drawn back, her coarse facial whiskers projected forward, her mouth held partly open as though to show me the large and powerful teeth that she wouldn't hesitate to use if I interfered. One of the young appeared quite unconcerned, more intent on drinking than on inspecting me, even though it did give me a quick glance out of its beady little eyes. The second cub was of a more nervous disposition; it came forward, stopped, stared at me, and growled, peeling back its lips to show me its already formidable fangs;

but it backed away until its hindquarters were partly concealed by the brush from which the family had emerged. The female, meanwhile, continued to eye me, standing at the water's edge, head raised, dark eyes fixed on mine. Then she began to drink.

The group was fairly typical of the species. Each wore a coat of long, coarse guard hairs concealed beneath which I knew was a second coat of dense, fine underfur liberally anointed with oils similar to the lanolin of sheep, which made it impervious to moisture and offered wonderful insulation. Although shadings were variable, each animal was similarly colored and patterned: The general tone of the coat was dark brown, a rich and glossy tint that captured the sunlight and turned almost auburn at certain angles; high along the forehead and at the tips of the ears the fur was gray, a sort of salt-and-pepper tinge; a band of light buff hair extended along each flank, broadcast over the shoulders, and ran backward to the short, stubby tail, where it merged with the band that traveled along the opposite flank. Irregularly shaped white markings were present on the chest and throat of each wolverine; the undersides were also dark brown.

The female was about 36 inches from nose to tail; the larger of the two cubs—the more timid of which I judged to be a male—was probably 26 or 27 inches long, his sister some 4 inches shorter.

Their bodies were chunky and obviously muscular; the tails, resembling in shape those of porcupines, were short and quite bushy, the mother's about 8 inches long, that of her cubs 1 or 2 inches shorter. The heads were broad, the snouts blunt, the eyes black. Below each rather massive body the legs appeared short; this was an optical illusion created by posture, for the wolverine moves like its relative, the weasel, bounding along in a somewhat hunched way, and even when standing, it seems to hold itself low to the ground. The feet of wolverine are similar to those of the black bear, but proportionately smaller; they differ in one important aspect: The five claws on each foot of the wolverine are partly retractile (they can be half-sheathed), thus protecting the sharp points from wear, while the claws of a bear are non-retractile.

Campfire tales have exaggerated all aspects of the wolverine's behavior. In reality, this little-understood animal is not nearly the savage that it is reported to be. Strong it is! It may well be the strongest predator, pound for pound, found in North America. Extremely fast and determined when attacking or escaping, it is also relentless and shows remarkable courage when defending itself, its young, or its food.

I was not in the least surprised by the audacity of the female who had dared to bring her young to drink in my presence; neither was I surprised by her calm, for I had observed similar behavior displayed by a number of other wolverines. Undoubtedly these chunky hunters are daring, but they are also intelligent and do not go out of their way to look for trouble if it can be avoided, tall tales notwithstanding. Only the young male displayed a measure of hostility; he did so because he was afraid. And even *he* eventually came to drink, undoubtedly gaining confidence from his mother and his more daring sister.

A fascinating characteristic of wolverines involves the way in which females conceive. They enter into a period of heat only once a year, but not necessarily at the same time each year. Males, on the other hand, become sexually active in April and remain in this condition until early September. But no matter when a female mates, her fertilized eggs do not become implanted in the uterine wall until the following

January,* when each egg continues normal development until the young are born in late March or mid-April, the entire gestation period lasting between seven and eight months, depending on when the mating took place. Such arrested embryonic development is not unusual in nature. In the wolverine and some other animals, it has evidently evolved in order to ensure the survival of both mother and young during hard winters, when an animal with rapid metabolism can only just manage to sustain itself and would probably not survive if it had to nourish from two to five developing embryos.

The three wolverines spent almost four minutes beside Spencer Bill Creek; then, led by the mother, who cast one last penetrating look in my direction, they lumbered back into the forest, leaving in their wake a slight but penetrating odor of musk.

After a noon break, I continued surveying the north side of the valley, sticking rigidly to my schedule despite many temptations to linger in particular locations or to follow promising tracks. In this way I progressed steadily toward the east, paddling, walking, charting, observing, and making copious notes.

Before the day was over I had reason to revise my estimates of the prey population, at least in the area where I was then working. I had thought that mule deer would be relatively

*In humans, for instance, a newly fertilized egg—at this stage called a zygote— begins cell division as it starts to travel down the oviduct preparing to complete a journey that will last between eight and ten days and during which its cellular divisions continue to increase. At the end of its migration, the zygote releases substances that destroy a small portion of the uterine lining and now—as an embryo—it becomes implanted in the wall of the uterus. During its migration, the zygote is kept nourished by a substance released by the maternal glands; afterward, as an embryo, it is linked to its mother's blood supply. Wolverine zygotes, however, take months to become implanted, an arrested development during which they are also nourished by the glandular secretions while cell development is interrupted.

abundant here, but I found that they were more scarce than plentiful. I had noted a number of their tracks, but all the indications were that these had been made by individual animals. Moose and mountain caribou, on the other hand, were clearly present in fair numbers. The first indication of this came soon after I had paddled away from Spencer Bill Creek and heard loud crashing noises coming from the south shore of the Goldstream. I turned my head to look in time to see three moose galloping for the cover of the forest. One was a magnificent bull with a huge rack of antlers, obviously in peak breeding condition; then there were a mature cow and a smaller female, probably a yearling. Each animal was in a different locale, but all had turned to run as the canoe rounded a bend. I had evidently interrupted the preliminaries of mating, for these large deer enter the rutting season in late September and continue in that condition until about mid-October.

Later in the afternoon I saw five more moose, each on its own, for when they are not engaged in the breeding rituals—or, if cows, are busy with their calves—these majestic animals are predominantly solitary. Three times I heard cows calling, a loud, bovine bellow ending in a harsh, coughing bark that echoed through the wilderness.

The moose appeared to be more numerous on the south bank of the river, but the north bank area yielded caribou. The first of these large ungulates that I saw were bunched in a wide part of the valley about half a mile from the river, sixteen animals comprising the group. Later, on shore, I found numerous tracks, droppings, and bedding places.

The next day, as well as the one after that, I saw several more moose on both sides of the river and a total of nineteen caribou in three groups; the first consisted of three animals, the next of nine, the last, seven. All appeared to be adults.

By the end of the fourth day, as I was searching for a

bivouac site at the end of the valley, I realized that I had underestimated the time that it would take me to survey the Goldstream valley. Up to now I had just done the north bank. In the morning I would turn around and retrace my course while exploring the south bank and its environs. This meant that I was going to have to put myself on short rations until I reached French Creek again, when I proposed to return to my base and pick up another six days' worth of supplies before tackling the north and south banks of the river and of the valley land west of French Creek. This area offers a great deal more flat land and eventually meets Goldstream Creek, which emerges from the beaver pond where I saw the owl, and then flows in a southwesterly direction to empty itself in the Columbia about four miles south of the mouth of the Goldstream River. Day after day I continued my survey, meeting a variety of animals, mapping the land, tracking, preparing another lean-to shelter.

On the sixth day, soon after I began to explore the western reaches of the river, I found a second set of puma tracks. These were smaller, the front imprints measuring $4^1/_4$ inches wide by $3^3/_4$ inches long. Were these the marks of a female lion? The next day I found a kill and more cat tracks, evidently made by the same animal. Within a section of brushland and amid the dead branches of a fallen spruce lay the remains of a relatively small caribou. It was an old kill, made perhaps four or five days before; a number of scavengers had eaten from it. The head, though already putrefying, was more or less intact, so I split it open in order to examine the animal's respiratory tract. I found that the caribou had been infested with bot-fly larvae, its nostrils, sinuses, and throat full of the obscene, maggotlike creatures, now dead, but still hooked to the mucus linings.

By now the weather was getting appreciably colder. Rain had fallen intermittently for several days and the snow line was moving downward steadily. I had to hurry, for already ice was forming on the ponds and on small pools as well as on the edges of the river.

By the evening of the eighth day I had reached Goldstream Creek and had explored its narrow valley up to the midway point, a distance of about two miles from the pond. It snowed that evening, a light, steady fall. When I awakened the next morning, the wilderness had been transformed. The sun was just showing behind the mountains as I emerged from the snow-laden tent to find an azure sky, unmarred by clouds, and serried ranks of freshly decorated evergreens. The temperature had dropped considerably overnight. I guessed that it was about twenty degrees Fahrenheit, the kind of cold that, under such skies as I was then enjoying, is more invigorating than uncomfortable. Still, I had to hurry. The north bank remained to be explored on my return journey to French Creek, and I had to keep ahead of the ice.

Nevertheless, I became so exhilarated by the transformed surroundings that immediately after I had breakfasted and taken down the tent, I stole two hours during which I just rambled through the wilderness.

I felt somewhat guilty about my frivolous behavior, but as I neared the beaver pond, I found a fresh set of lion tracks and so justified my lapse. The puma had descended from somewhere up the flanks of a small round mountain (3,700 feet at the peak), slaked its thirst at the pond, and then traveled up the Goldstream Creek valley. Fresh snow had partly covered the animal's marks, so I had no means of determining when it had passed this way, except that it must have done so between last night, after the snow had begun to fall, and early morning, not long before the weather cleared and precipitation ceased.

For two more days I enjoyed fine, crisp weather as I surveyed the northwest shorelines of the river on my way back to French Creek, but then the sky clouded over and more snow began to fall—big, heavy flakes that floated straight down, for there was hardly any wind. But later the skies became darker and darker as the flakes grew in number and size. I was in for a nasty time!

I was then opposite the mouth of McCullock Creek and about six or seven river miles away from French Creek, but because of the weather I decided to forgo further explorations and head for home.

It was just as well that I did. An hour later I was almost blinded by a thick white blanket now being driven by an east wind that shrieked its demented way along the valley floor. Paddling against this blizzard became an almost unendurable ordeal. My hands ached from the cold; my eyes were constantly stung by the driving snow, despite the fact that I slitted them to the point where I could hardly see.

It took me almost four hours to cover those seven miles of river and almost an hour beyond that time to drag the canoe ashore and stagger uptrail to the valley of the shack. Never was I more thankful to sit before a roaring stove while sipping hot coffee laced with a dram of scotch whiskey. Outside, the storm raged on.

THE SOMEWHAT UNUSUAL AUTUMNAL BLIZ-
zard lasted for eighteen hours and dumped
nine inches of snow on the bottomlands,
but considerably more in those places where
it had become drifted by the northeast wind
that spawned the storm. For me, warm and
comfortable in my small shack, the time went swiftly as I
occupied myself in expanding the notes made during my
survey, now and then interrupting this task to go outside to
get firewood, or to stand beside my shelter, absorbing the
feral qualities of the elements.

After I had coordinated my findings with the map, marking
on it geographical, topographical, and botanical details of
importance, I listed the animals observed and keyed each
sighting with the chart. Altogether I had seen sixty-seven
woodland caribou in seven bands that ranged over various
sections of the river, from the east end of the valley to the
west end, where it opened into the narrow Goldstream Creek
gorge. I did not actually sight more moose than those noted
upriver, but I heard a number of them call and several times
listened to their crashing sounds as they fled my presence. I
also saw eight mule deer, each animal in a different location.

While working my way downriver, somewhere about midway between the eastern end of the valley and the mouth of
French Creek, I met an enormous male grizzly, a fat but
majestic animal whose coat shone in the sunlight. He was
grumbling audibly as he descended a steep, rocky area north
of the river while I was walking on the relatively flat land
below. Since he did not appear inclined to alter his course
away from mine, I obliged by moving out of his line of travel.
Standing near a slim but branchy spruce, the limbs of which
extended almost to the ground, I watched as the big bear
scrambled down to the flatland at a point about 150 feet from
where I had taken up my station beside the easy-to-climb

tree. But such precaution was unnecessary. The grizzly cut straight across the north side of the valley, squelched through the marsh area, and flopped himself into the water, his "bow wave" big enough to have swamped a canoe, had it been floating near the great animal. I watched as he swam across the stream, landed on the other side, shook gallons of water from his shaggy coat, and then scrambled away into the forest. Not once did he dignify my presence by looking my way, yet he undoubtedly knew that I was in his neighborhood. I later saw another of his kind in the western end of the valley, a smaller male who grunted as he wheeled away from the riverbank when my canoe rounded a bend. This one left hurriedly and was soon lost among the trees and shrubs.

Black bears were evidently plentiful in the area, for I saw five in various places, including a mother and two yearling cubs that might have been the same group I had seen during my first upriver journey. I also saw two billy goats, but from a distance. Snowshoe hares were abundant, undoubtedly enjoying one of their many population peaks; so were brown lemmings, ruffed grouse, and ptarmigan.

On three occasions I stopped to fish and was each time rewarded with fat rainbow trout, weighing about two pounds each. The fish were most welcome as a change of diet and boded well for future supplies in the spring, when I would undoubtedly find myself in need of fresh protein.

During the first night of the storm and late into the next afternoon, the wind was persistent and furious, howling down from the peaks, whistling shrilly through the trees, swirling snow in all directions. But the precipitation was intermittent, at times stopping altogether, at other times falling in tiny, sugarlike particles, on yet other occasions arriving as big flakes that, within the shelter of trees, appeared to be the size of coat buttons. Once, at first light the next morning, I watched the small clearing as cat's-paws of wind tossed and swirled the snow in all directions, reducing visibility to mere yards despite the fact that new snow was not falling. I had thought about going out for a time, but this display caused me to change my mind. There was no sign of my three friends, Wisa, Ked, and Jak. They had no doubt elected to sit out the storm secure in some evergreen whose thick lattice of branches and needles gave them all the shelter and comfort that these tough northern birds require. My thoughts then turned toward the puma. Like all the other animals of the forest, I felt sure that it too had taken shelter. It might well be hungry, and anxious for the storm to end, but it would not otherwise be too discommoded.

The wind stopped blowing at about 3:00 P.M., but the skies remained overcast, wrapping the forest in shadows. I went outside and began to walk toward the collapsed mine building, but before I was halfway there the three gray jays found me. Ked landed on my hat, Jak on my left shoulder, and Wisa, always the great opportunist, hovered in front of my face, her wings beating themselves into a blur as she waited for my hand to emerge from the pocket into which it had disappeared when I saw the trio approach. I had by this time made it a point always to carry a supply of mixed nuts and oatmeal for just such occasions, so when I opened my palm, it was Wisa who landed on it first; but she was soon followed by Ked and Jak. For some minutes I stood there, a living bird feeder, while the jays first had a good meal—they were obviously very hungry—and then made a number of trips with full beaks, hiding their spoils and returning for more. After five or six of these flights I moved on, but kept a supply of food in my hand, ready for each of the birds as it came to gather more rations.

I spent about fifteen minutes walking around the small clearing, but apart from the jays, there was no other sign of life, not even tracks of hares and mice. This was not unusual. The animals of the wilderness are not as a rule in a hurry to emerge on the heels of a violent storm. In any event, at this season, the hour at which peace arrived coincided with the time when daytime animals are getting ready to bed down and nighttime animals are not yet due to emerge in quest of food. I was just thinking that it would be unusual to find new tracks in the snow, or to hear animal sounds, when a pack of timber wolves began to howl. The primordial, haunting chorus was coming from the southeast, upriver, perhaps in the vicinity of Spencer Bill Creek. The pack howled continuously for about five minutes, their ululating voices echoing through the valley in a wild and magnificent fugue that ended all too soon.

Before going to bed that night, I read for a time, choosing Thoreau's *Walden*, which I turn to on those occasions when I am pensive and seeking to unravel some of my own ideas about nature. I can't say that I always agree with Thoreau, but there is in his work much that stimulates. My reading was intermittently interrupted by periods of meditation. Then, not long before I turned out the kerosene lantern, one particular passage struck me as being especially pertinent to my present quest:

> We need the tonic of the wildness, to wade sometimes in marshes where the bittern and the meadow-hens lurk, and hear the booming of the snipe; to smell the whispering sedge where only some wilder and more solitary fowl builds her nest, and the mink crawls with its belly close to the ground. At the same time we are earnest to explore and learn all things, we require that all things be mysterious and unexplorable, that land and sea be infinitely wild and unsurveyed and unfathomed by us because unfathomable.

I had come here from the unfathomable sea to seek my own particular tonic in the wildness; I had already waded in marshes, and although the birds and sounds and animals here were somewhat different from those that Thoreau was thinking about when he sat down to make an entry in his diary, I understood his meaning. I was "earnest to explore and learn," particularly about the puma; and I also required the mystery to continue forever, for without it I knew that I would find life very dull.

As I lay in the darkness, listening to the companionable chatter of logs burning in the stove, my mind journeyed into the wilderness night and saw the puma tracking through the unblemished snow in quest of food, and watched it eating of its kill, then lying replete in casual sanctuary while lazily cleaning the blood from its fur, a beautifully sleek, magnificently wild creation whose amber eyes returned my gaze with a look of recognition that gave us understanding of each other; we were friends who had never met, but who assuredly one day would come together in peace and tolerance.

The morning after the storm, I awakened as the sun was already edging up to the highest peaks of the eastern mountains. This meant that I had lost about two hours of light at the start of a day during which I planned to be extremely busy.

Dawn is a somewhat somber event when observed from within an alpine valley, for the sun must rise above the summits before it can reach the bottomlands. Even after the full solar display has cleared the tallest crests, those slopes that face west remain in shadow until almost noon, when the sun has advanced to a point close to its central zenith.

In the Western Cordilleras, that awesome chain of upthrust granite that stretches from Alaska all the way down to Cape Horn, the southernmost tip of South America, dawn is a muted occurrence.

Sunup in a mountain valley is first seen as a bluish hue that brings into relief the upper third of the serried crests, a cobalt wash that contrasts with the surrounding darkness, lingers for a time, becomes noticeably paler, and is streaked with phosphorescent green. As the pinnacles of each mountain become visible, pinkish overtones emerge and mix with the blue-green of first light. Slowly the pink intensifies. Each peak becomes backlighted and soon the eastern skyline is streaked by shafts of orange. On the ground, the transformation from night to day resembles the dusk in other places, when, after the last rays of sun have gone, twilight deepens until it becomes penumbral. The opposite takes place in alpine bottomlands, of course, penumbra giving way to twilight and, as the sun climbs higher, being replaced by daylight.

Waking late, I missed the kaleidoscopic effects of the dawn, but I was in time to see the sun begin to edge up the tallest peaks. Irritated, for I had intended to rise at 5:00 A.M., I looked at the small traveling alarm clock that I had forgotten to set the night before and saw that it was seven-thirty; I also noted that the shack was cold and that the fire had gone out. But, being used to chilly awakenings, I ignored the cold, got out of bed, and, dressed only in undershorts, walked to the door. I was anxious to see the day, now that the skies were clear and the landscape was mantled in fresh white.

From the open doorway a wonder-world greeted me. Sculptures of snow decorated every tree in sight; some shapes were small and simple, others large and intricate. There were thousands of natural art forms of a surrealistic quality beyond the craft of man. The contrast between the white and the dark of trunks and branches was bold and varied: Some trees were wet, their boles olive green; others were chestnut-brown. Sprigs of bright green needles wore diadems of crystalline snow; branches were decorated with sinuous ropes of white that wound around their supports, here and there breaking and loosening small cascades of glittering powder. Each of last spring's plants and grasses was individually made splendid by wisps of white, or by pristine buttons resembling freshly confected marshmallows. The ground was entirely concealed by an alabaster mantle that followed the contours, dipping here, rising there, climbing boulders and draping them, but leaving small, uncovered openings in places, like darkened windows through which fronds of dried ferns, lichens, and mosses protruded.

Several snowshoe hares had passed within eighteen inches of my doorway, as had a red fox. As I looked at the paradisiacal scene, Wisa, Ked, and Jak flew to me. One bird landed on my head, and each of the others selected a shoulder to perch on, their claws dimpling my bare skin. Moving slowly, I backed into the shack. The birds stayed with me. Inside, the trio flapped down to the table, and as I was getting a handful of trail rations they helped themselves to some mixed nuts that remained in a tin cup from which I had been munching the night before. When I returned with the food, the jays abandoned the salty leftovers in favor of their preferred treat, each swallowing busily for some moments, then cramming their beaks and gullets with the bait before flying out the door to hoard it in the forest.

I closed the door, dressed, and started a fire. I was suddenly impatient to go out into the transformed wilderness and to continue the explorations of the puma range that had been interrupted by the blizzard. While the coffee water was heating, I went to the icy stream with a basin, a towel, and a bar of soap, washed *al fresco,* and hurried back inside to stand in front of the stove until its warmth restored the circulation in my hands and face. Then I breakfasted.

An hour later, dressed for the trail and carrying a packsack that contained enough food to last thirty-six hours, my down sleeping bag, and an axe, I left the shack, securing the door and window shutters. Strapping my heavy bush knife around my waist, I walked down to French Creek, where I noticed that the snow was soft and fluffy and about six inches deep wherever it had not been blown into drifts. As I had surmised before leaving my shelter, I would not yet need snowshoes to travel the land.

Around me, chickadees and nuthatches were calling almost continuously, each tough little bird hunting for dormant insects within the branches of the trees. Woodpeckers were also busy: the big pileateds with the blood-red crests, the medium hairies, and the small downies. They called shrilly from time to time and pounded dead trees as they probed for insects, the pileated's staccato drumbeats outclassing the more subdued tapping of the smaller members of the avian family. Now and then one would fly across the valley, progressing with the up-and-down motion characteristic of all woodpeckers, which results from their habit of beating six times, then folding their wings for the duration of one stroke before they flap again, a rhythm endlessly repeated that causes the birds to drop slightly each time their wings rest and to rise again to the former level when they resume flight.

By eleven-thirty that morning the sun flooded down into the French Creek valley, beaming out of an azure sky through which scudded a few small and milk-white clouds, the golden light and the sparkling snow transforming the land, turning it into an ethereal park through which I walked with a sense of excitement and pleasurable anticipation. I had already noted a great many animal tracks, including those of deer and moose. Indeed, two moose had wandered across the valley where I had built my shack, pausing here and there to browse off the young poplar and birch tips, seeking the buds as well as the chewy stems with the tender bark that tastes like fresh cucumber.

Soon after descending to the creekbank, I saw a solitary wolverine, a large male who galloped away from a place where the snow was trampled and showed scarlet. Going to look, I found the few remains of a snowshoe hare, mostly bits of bone and sinew and clumps of white fur. Tracks leading to the site told their own story: The hare had been running, but the wolverine had intercepted, killed, and eaten it. The action must have taken place fifteen or twenty minutes before my arrival, for the blood, though clotted, was not yet frozen and was a fresh red color.

Since I was continuing the survey of this region, I forced myself to examine each sign of life as I came upon it, resisting the urge to take shortcuts in the hope of finding puma tracks. As I had done along both banks of the Goldstream, so I was doing now, except that I was not, of course, using the canoe. My plan was to explore the west bank of French Creek to the end of the valley through which the waterway coursed, a distance of about nine miles. Pausing as frequently as I was, I knew that the survey would take all day, so I had come prepared to spend the night at the end of the valley, and to build a bivouac shelter that would remain for later use. In the morning I intended to examine the east bank, working my way back to home base.

Counting and classifying tracks and marking their locations on a map is a tedious business if it is done properly, but it is also a useful one because it builds an accurate picture of the life of the wilderness while supplying details of vegetation and topography that no survey chart is able to show.

By late afternoon, walking in shadow because the sun had disappeared behind the western peaks, but still under clear skies and moderating temperatures, I reached that part of the valley where marshland, ponds, and small lakes predominate on both sides of the river. Here, for four miles, the bottom is wider, but exceptionally wet, fed by seventeen small creeks. Compared with the first half of the valley, vegetation differs radically in this area. Now I encountered red cedars—tall, thick trees that grew singly, rather than in dense clumps, quite spectacular plants that are counted among the oldest in Canada, the growth rings of some of them showing that they have been alive for more than eight hundred years. The biggest and tallest members of this species are found mainly in coastal regions. (One of these, cut down in 1948, was 200 feet tall and had a trunk diameter of more than 13 feet, giving it a circumference of more than 40 feet.) Those I encountered that afternoon were not giants, yet they reached about 120 feet in height and had diameters of between 4 and 5 feet.

Below the cedars, in boggier ground, grew Labrador tea, scrub birches, willows, and mountain alders in conditions so crowded that the swamp showed no openings. With visibility restricted, I climbed higher up the west bank until I could look down into the marsh area. Almost immediately I saw seven caribou, a small band that was loosely clustered near the east shore, munching on the shrubs and marsh grasses. I had intended to slip past the group, not wishing to disturb the animals, but on seeking to go a little higher and gain the shelter of a dense stand of hemlocks, I reached to grasp a shrubby plant to aid my ascent. What seemed like thousands of needles stabbed into my fingers and palm, and in my hurry to let go, I slipped, dislodging a number of loose rocks that rattled downslope. By the time I regained my footing the caribou had gone, though I wasn't greatly preoccupied with them just then. I was too busy inspecting my hand and cursing my stupidity for failing to recognize an upright stem of devil's club (*Oplopanax horridus*), a fearsomely prickly plant with leaves—in season—like the ears of a small elephant that, like the stems and branchlets, are covered in strong, pale yellow, extremely sharp thorns. I spent the next fifteen minutes plucking the pins out of my hand, an operation that would have been impossible without the pair of tweezers that I always carry in a belt first-aid kit whenever I am in the wilderness.

I reached the end of the valley within about half an hour of dusk, and there, on a relatively high, dry section of land on the east bank, I unloaded the pack and began erecting a pole-and-brush bivouac, a structure seven feet long, four feet high at the front, and two feet high at the rear. The roof poles sloped abruptly in order to shed moisture, should there be any. Before I piled evergreen boughs on top of the make-shift rafters, I draped them with a sheet of heavy-duty poly-ethylene. When I had stacked the evergreen boughs on top of the shelter six inches deep, I did the same to the sides of the lean-to, sealing its ends against the wind. The structure's opening faced south.

Five feet in front of the bivouac I constructed a reflector wall eight feet long, so that it overlapped the shelter a foot on either side. This parapet was built from poles leaning toward the structure and resting on a frame consisting of two six-foot-high uprights united by an eight-foot-long crosspiece; behind the wall poles I also piled evergreen boughs.

Between the reflector shield and the bivouac I next built what I call a long fire, that is, a fire six feet in length and about fourteen inches in width. First I cut two large, dry logs with a diameter between eight and fourteen inches and about six feet long, which I then set lengthwise on the ground, fourteen inches apart. Within this space I stacked smaller logs, also about six feet long, the thickest at the bottom, the smallest pyramiding toward the top. Then I placed a six-foot line of kindling on the last layer of logs. Once I had lit the kindling in about six places along the line, the flames united and the fire burned from the top down, producing an even, relatively slow-burning heat source that would last all night without the need to add fresh fuel. Such a fire distributes heat along the entire length of the shelter.

Working methodically, it takes about an hour to build this kind of camp—provided one has selected a site that offers a good, accessible supply of building materials and fuel wood—and, of course, it becomes a permanent shelter, usable without more effort than gathering fresh firewood on those occasions when one needs to sleep out in the same general area. For one-night stops in other places there are different, faster ways of keeping warm, but these are not as comfortable or secure.

It was dark by the time I had settled myself within the bivouac, relaxing while my supper cooked on the fire. Later, after I had eaten, I spent some time bringing the map up to date, my light coming partly from the fire and partly from a one-pint glass jam jar in which I had stuck a candle. This is my usual, simple night-light when camping this way. Apart from providing a reasonable amount of light, it can also be used in a snow shelter to keep the inside temperature above the freezing mark. As a safety precaution, I usually put two inches of water in the bottom of the jar, so that if I fall asleep while the candle is burning, the water will put out the flame when the level of wax reaches it.

The next morning I was up ahead of the dawn and had eaten breakfast and packed my equipment even before the first cobalt flush appeared in the sky. The valley and its surrounding forest were quite dark, since there was no moon that night, but as I sat enjoying my first pipe of the day—and making it last because I would not smoke again until the noon break—I listened to the sounds that came from all around me, seeking to identify their makers.

Mice scurried under the snow, sometimes emerging on the surface to look for seeds. Their progress was muted when they ran through their snow tunnels, but was audible if they were nearby. When they emerged, they were relatively noisy because they scrambled up the stems of dead plants and bushes. Down the valley, a great horned owl was calling, its deep voice magnified by the mountain echoes; across on the east side of the valley, but somewhere up the flanks of a mountain, a large animal was moving; I guessed it was a moose.

As I sat watching, the shadows became faint and the trees developed outlines. I began to hear chickadees calling quietly, still sitting on their roosting perches, but preparing to move into the forest in quest of their breakfasts of hibernating insects. Then I heard the wolves howl. They were distant and to the south, but their haunting song floated up the valley, muted, yet as fascinating as ever. For almost five minutes the pack howled, sometimes in unison, on other occasions singly or in twos and threes, indulging in one of the impromptu singsongs to which wolves are prone and which they appear to enjoy for the companionship that it brings them. Perhaps it is a sense of union, of closeness as a society.

In the subdued light of new day I left the shelter and began to explore the west bank, walking slowly, having to bend often to examine tracks or to avoid sharp, dry branchlets that could easily put out an eye if I blundered into them. In this way I progressed at a rate of no more than a mile an hour until the light intensified; then I managed about two miles an hour, slowing to observe the profusion of tracks that I encountered all along the edges of French Creek and up the flanks of the mountains. Grizzly and black bears, moose, deer, caribou, foxes, coyotes, wolverines, and hares were some of the animals that left a clear record in the snow. But I saw neither puma nor wolf tracks that morning. The bear tracks startled me at first, for I would have expected these animals to be holed up for the winter; but on reflection I realized that the snow was light and the weather was not quite cold enough to drive them into their dens.

At noon, after six hours of travel, I stopped for lunch at the end of the marsh area, near the last of the tiny lakes. Munching trail rations and sipping water, I sat quietly, partly in sunshine and partly in the shadows of the trees. I was just thinking about reaching for my pipe when a shower of snow descended from a tall hemlock that grew two hundred yards east-southeast from where I sat. Reaching slowly and carefully for the field glasses while otherwise keeping absolutely still, I held my eyes on the tree; more snow was falling and some of the branches were shaking vigorously. With infinite care I raised the glasses to my eyes, adjusting the focusing wheel until they were in sharp focus. Seconds later I saw it. The puma at last!

Only the big, broad head and part of the chest were as yet visible, but the shape of the pate, and particularly its width, made me certain that it was a male. Its visage reminded me poignantly of my friend Tom.

My excitement was intense, causing my heart to beat so loudly that I fancied the puma would hear it from its perch. But I remained immobile, watching. For some moments the big cat did not move, its yellow eyes scanning the country below, not looking in my direction. Then, perhaps a minute after I first saw it, the cougar withdrew its head and forequarters from the foliage, and more snow began to fall, but lower down the tree; the cat was descending. I continued to keep the glasses fixed on the hemlock and watched as the telltale cascades of snow descended the tree, allowing me to follow the unseen puma's progress. About thirty feet from the ground the branches were thinner, some dead. Here I saw the entire cat when it turned broadside to me. It was big, probably measuring close to 8 feet from end of nose to tip of tail, and it was heavy. Now it turned away from me, dropping its front paws to a lower branch. As it did so, it cocked its rear up in the air and its tail moved sideways, allowing me to determine its sex. It *was* a tom.

The cougar surveyed the land below from its unsteady perch, at first staring fixedly southward while its ears moved, flicking spasmodically as they searched the environment for sounds. Soon it turned its head to the left, in my direction, and I was able to note that while its ears sought to pick up noise, its nostrils were twitching continuously, seeking odors. In this way, as systematically as a radar dish, it continued to probe the wilderness, quartering the territory by moving its head, keeping it in one position for some seconds while its ears and nose worked vigorously, then directing its faculties toward a new section of forest. It was also using its eyes, of course, but given the nature of the terrain, which was thickly covered with trees and bushes, vision was playing a minor role, subservient to smell and hearing.

After inspecting the wilderness that lay to the south, west, and east within a rough 180-degree arc, the tom turned, scrambling so that it reversed itself, its forequarters now supported by the upper branch and its hindquarters positioned on the lower perch. In this posture it again probed the forest, turning its head to the left, in a northwesterly direction, then, quadrant by quadrant, moving in stages to the right, toward me and the southeast.

I found myself in a quandary. My position was somewhat above the level where the hemlock grew, and there was an irregular corridor of open land between me and the cat; if I kept the glasses focused on him, the glint of their lenses would almost certainly attract his notice; but if I lowered them, even the most gradual movement would be detected by the cat's peripheral vision, which, as in all mammals, is insensitive to color but especially sensitive to movement and light. Indeed, peripheral vision is probably the most important sight faculty for all wild animals (and for man also, when he lives in the wilderness) because the ability to see movement outside of the limited focusing range of the pupils alerts the hunter to the presence of possible prey (or enemies) and, equally, warns the hunted of danger. The eyes of all hunting animals are positioned within the frontal part of the head, while those of the prey species are set on the sides. This placement allows the hunted a considerable advantage over the predator, for it extends the field of peripheral vision to almost 180 degrees for each eye as opposed to something less than 90 degrees for predators. On the other hand, hunters have much keener binocular vision, and this allows them to detect three-dimensional outlines. On balance, however, it seems to me that the prey animals have a visual edge over the predators, a survival factor that helps *both* the herbivore and the carnivore: In the former case, the prey has a better chance of survival; in the latter, the hunter is forced to work hard in order to survive. This maintains the efficiency of the predatory species and, perhaps more important, prevents it from killing off all the prey in its territory.

The fundamental law of natural conservation decrees that the hunted may eat at will and with little conscious effort or planning within a limited range, the geography of which it knows intimately. To stay alive in the presence of bountiful supplies of food, members of the prey species only need to remain alert to the presence of danger and to know instantly the locations of their protective shelters. Conversely, the hunter must *learn* to find food over a far wider range, the size of which forbids the kind of geographic intimacy enjoyed by the quarry. Additionally, the predator must get to know the general habits of its favored prey species and even the specific habits of individual animals, learning these by trial and error until it succeeds in killing the quarry, or admits defeat and goes after a less alert or experienced individual.

This does not mean survival is always easy for the hunted and difficult for the hunter. On the contrary! There are many other factors and circumstances that play significant roles in the lives of all animals; but these were the basic differences between prey and predator that concerned me that day as I watched the puma.

The tom was surveying part of his range, just as I had been seeking signs that would help me to determine the kind and numbers of animals that lived in the region of French Creek while I especially looked for the marks left by the cougar. Now my knowledge of predator/prey relationships allowed me to recognize the dilemma I was in. I had not expected this encounter, so I had not placed myself in full shelter when I stopped for a break. It was only a matter of seconds before the puma would turn its attention to the quadrant of territory in which I sat, so I could lower the glasses slowly and hope to escape detection, or I could remain immobile and pray that the glint of the lenses would not give me away.

My concern and apprehension centered not on my safety, but rather on my eagerness to continue to observe the cat. If it detected me, I fully expected it to run, depriving me of a heaven-sent opportunity to study it.

As I continued to watch the tom, suddenly the need to make a decision about the field glasses was taken out of my hands. The puma turned his head and fixed his amber eyes on *me* as squarely as the glasses were fixed on *him*. The 7×35 magnification bridged the distance between us, cutting it down to about two hundred feet and giving me a good view of the whole animal. But it was his head, and especially his eyes, that commanded most of my attention as we stared at one another.

After some moments the movement of the tom's black-tipped tail attracted my notice. About four inches of the end twitched from side to side, a motion unlike the more vigorous lashing made when a feline is nervous, stalking prey, or annoyed. Tom, in London, had resorted to "tip wagging" when he was curious or puzzled, the jerking of the tail tip corresponding to a change in the position of his whiskers, which he jutted forward.

Shifting my gaze from the tail to the animal's muzzle, I saw that this cougar had also moved his whiskers from the normal, slightly swept-back position to an acute forward angle, each whisker curving slightly and protruding beyond the nose and lips. At the same time, the rounded ears were cupped in my direction and the nostrils were twitching actively. The tom's mouth was slightly open, the tip of his pink tongue visible between the lips.

We continued to watch each other, and I knew that he was aware of my presence. Yet he was neither seeking to escape nor showing signs of aggression. What would he do, I wondered, if I lowered the glasses? I was loath to interrupt my observation of the great cat, but on the premise that if I didn't act soon, he would, I decided to experiment.

I brought the glasses down rhythmically but not too slowly, the kind of motion that I knew from experience signals neutrality, a casual and self-assured change of posture that in no way resembles the stealthy approach of a stalking animal, nor the sudden alteration of stance or pace resulting from fear or aggression. As I lowered my arms, my eyes remained fixed on the puma, and although I could not now see his features clearly, his body, color, and pose were easily discernible across the six hundred feet that separated us.

The puma continued to watch me without showing signs of unrest. It lowered its head slightly, following the motion of my arms, and when I once more raised the glasses, it kept pace with my action. Now, since he gave no indication of alarm, or even of wanting to climb higher or to leave the tree altogether, I decided to try another gambit, lowering the glasses for the second time and at the same moment rising to my feet. The tom watched me, but stayed on his perch. I started to walk, sauntering down the gentle slope toward the eastern bank of the creek while keeping my head turned toward him; he was now watching me intently. In this way I covered about seventy-five yards during the course of ninety measured strides, traveling along a slightly diagonal course that kept the cougar's tree more or less in line with my right side. About fifty feet from the creekbank I stopped and squatted, raising my head to look directly at the animal. His ears were pricked forward, his head turned toward me; his stance suggested interest, or perhaps curiosity.

Because a treed cat feels itself secure, provided it is not angered or made uneasy by noisy dogs and equally rowdy hunters, I now reasoned that the tom would stay on his perch if I approached him more closely. He might, of course, climb higher, but I didn't think he would descend and run into the forest. At any rate, I decided to chance his escape, for I wanted to get nearer so that I could see him better with the naked eye and so that I might be able to talk to him quietly. But before standing up, I looked away from him, pretending interest in the wilderness and moving my head slowly across the vista, much as he had done when he was surveying his own domain. At least a minute passed as I swung my gaze southeast and returned it again to the northwest, until I was again able to focus on the lion. During this time he had continued to watch me, shifting his position only slightly. Now, ignoring him, I stood up, turned so that I faced north, and then started to walk along the creekbank. When I was level with his tree, I stopped, stretched slowly and deliberately, and looked for a suitable seat. A fallen tree offered a snowy perch, and I gave the lion my back and occupied myself with clearing the snow from the bark. When this was done, I began talking quietly, keeping my back to him for some moments before turning slowly and sitting down on the deadfall hemlock. Now I allowed myself to raise my head to look at him, knowing, as my gaze traveled up the hemlock's trunk, that I would see him still poised on the same two branches, for not even a puma can move soundlessly up or down a dry, scaly-barked tree. From my position almost at the water's edge, I estimated that only seventy-five feet lay between me and the tom, and I found myself looking directly into his golden eyes, simultaneously realizing that I could see every detail of the great, sinuous body. I had not stopped my soft-voiced monologue, and now, able to observe even the smallest change of expression on his face, I knew that he was listening to me with the same kind of interest exhibited by my old friend in Regent's Park Zoo.

Not wishing to stare fixedly at him, because animals become uneasy under such intent scrutiny, I kept moving my eyes, occasionally turning my head to look at other objects in the vicinity. During one of these changes of objective, the tom began to move. I heard the quiet scratching sound he made as he shifted his stance. When I turned my face toward him, I saw that he had settled himself down on the two branches, his hind end resting on his haunches, his chest on the upper perch, one of his great forepaws half-cupped around the branch, the other hanging limply. His ears were held midway between the vertical and the horizontal—what I call the "half-mast" position that in all wild animals signifies relaxation. His long tail hung straight down, its black tip concealed among the smaller branches and needles. Then he yawned, a prodigious gape during which his mouth opened wide, his eyes squinted shut, and the skin on his muzzle and on either side of his nose wrinkled as though from sudden age. The rough, pink tongue arched inside the mouth, revealing all the teeth and the interior of the cavern glistening with saliva.

We had made contact! I cannot fully describe the thrill that I felt on realizing that this wild and magnificent animal was showing me that he was prepared to trust me. I was not now excited; instead, I felt a glow of intense well-being, a sense of calm pleasure so exquisite that it bordered on the sensual. And I realized that I was undergoing a distinct personality change, a mind division that enabled my cortex to continue to take note of and to log the messages of the environment that reached it through my perceptual faculties, while at the same time my subconscious senses became acutely active, giving me an awareness and insight of preternatural proportions. This was not a new experience. I had first become conscious of it years before, while living entirely alone in the wilderness.

At the time I had been astounded by the discovery of what I considered to be a second self, as though there were actually two distinct and different entities sharing the same mind: the one a modern, civilized pragmatist whose philosophy only accepted measurable values that would yield positive results after due analysis; the other a young and inexperienced mystic, a sort of adolescent seer gifted with precognitive powers, but unable to employ them on command. Later, realizing that I had begun to develop and use an ever-present, latent aspect of my personality, I was content to drift in and out of the state when I was gripped by it, thinking that the condition represented both my clinical and my human self. As the years passed, I became so accustomed to these periods of duality that I began to accept them without question, enjoying them and deriving value from them when they occurred, but not deliberately seeking to create them at will. Vaguely, I recognized that I was developing extrasensory powers, both in the matter of perception *and* transmission, but inasmuch as

I did not understand the forces that governed such phenomena, I didn't seek to analyze them. Indeed, if anything, I was somewhat skeptical, fearing that I was deluding myself into believing in something that could not really exist.

During the early 1960s I had an opportunity to attend a symposium on parapsychology and psychokinesis at Duke University, in Durham, North Carolina, during the course of which I realized for the first time that extrasensory communication is both real and far from rare. The researchers attending the conference did not seek to explain the phenomena under discussion, and the whole affair was as dry and undramatic as the scientists could possibly make it. The reason for such a dispassionate approach was that Duke had taken the plunge into a subject that had until then been considered disreputable within the exalted halls of science. But during a series of private conversations with researchers from a number of universities in North America and Europe, I learned that all of them were confident that mind communication was an art that had become lost to mankind because of the influences of civilization. Many of the scientists present believed that the forces responsible for the faculty would one day be understood and harnessed, making it possible for people to communicate at will over long distances without electronic aids.

My experience at Duke caused me to reexamine the many relationships that I had developed with wild animals, when I had somehow understood their needs and moods while managing to convey to them that I was a being that would help them and could be trusted.

Afterward, I sought to harness my extrasensory abilities, to use them at will. But I could not. Again and again I would try hard to win the confidence of an animal, only to find that it ran away from me. Conversely, when I least expected it, and usually when I found an animal that was in need of care, I always managed to reach it in some way. As the years passed, I gave up the attempt to command ESP to appear on demand and I allowed my second mind-self to work unfettered, popping up when it felt like doing so. Significantly, it always did so when the situation was most critical for myself or for the animal in which I was interested.

Satisfied with the status quo, I thought little about the matter, accepting it as a useful adjunct to my work with wild animals (and sometimes with people), but then the whale I called Klem entered my life and brought the whole subject into sharp focus, even though it was not until after the fact that I understood for the first time that I was capable of summoning an animal by means of extrasensory transmission.

The orca, a member of the so-called killer whale species, visited me for the first time while my boat, the *Stella Maris*, was anchored in an isolated inlet off the coast of northern British Columbia. I had been about to dissect a wolf-eel that I had unwittingly caught (and had to kill to get it off the hook without losing some fingers!) when the whale rose alongside the boat, the sound of his escaping breath alerting me to another presence nearby. Looking up from my task, I was startled to see the great mammal floating sedately on the surface and fixing me with one immobile, piercing eye. On impulse, I cut a two-foot chunk off the wolf-eel and tossed it to the whale, who immediately ingested it. That marked the start of a friendship that lasted several weeks, for Klem visited me at frequent and regular intervals.

One evening after my ocean research had ended and I had decided to study the puma, I found myself thinking about Klem.

Only then did it occur to me that, apart from the orca's first visit, he had almost always turned up within a short time after he entered my mind.

I had kept full notes, as I always do, and on checking them I saw that in seventeen out of twenty-three daily entries, I had mentioned that Klem arrived a short time after I had started to think about him continuously; that is to say, the orca was often on my mind during the period that spanned our friendship, but there were occasions when I thought about him *hard*, wondering what he and his family were doing, where they were, how they lived, whether he would visit me again. . . . Fortunately I had logged all of these thoughts, or at least most of them, for I could well have missed making notes on the subject on some occasions. I had also taken notice of the time lag between my thinking about Klem and his arrival. My notes verified that he came between fifteen and forty-five minutes after his image entered my mind.

That evening, sitting within the darkened mountains of southern British Columbia, I at last conceded to myself that ESP/EST existed, that it could be harnessed, and that I would try to use it in order to establish a close relationship with the puma that I hoped to find. Tom, in London, had managed to intercept my thought emissions and knew that I was coming to see him; Klem, in the Pacific Ocean, had picked up my mind signals and had come to visit. In a similar vein—though not nearly so well established—my postulate was confirmed by the many animals I had befriended, and the equally numerous wildlings I had met casually in the wilderness who had accepted my presence in their domain.

Now, sitting on the hemlock, watching the tom sprawled relaxed in his tree, I had no further doubts. I had communicated with the lion. Would he, in time, communicate with me? Or had he already done so when he showed himself to me? For all I knew, he could well have detected my presence before I was aware of him, then decided to show himself to me and to further tell me by his actions that he was willing to accept me on his range. Some people, I thought then, would consider such ideas pure anthropomorphism. As I sat looking up at the lion, the opinions of rigid-thinking individuals didn't matter to me. I had long ago decided that anthropomorphism wasn't such a great sin. Indeed, it is perhaps a good thing, for it allows a human to relate with an animal, to see it as a fellow being and to be willing to share the world with it. The approach to peaceful coexistence lies along that route, it seems to me.

FOR TWO HOURS I SAT ON THE LOG, RISING to move about occasionally so as not to become stiff, while the puma relaxed on his perch—though how an animal could manage to remain comfortable while draped like a sack of meal over two branches is beyond my understanding! He even managed to doze intermittently, breathing heavily and evenly, his muzzle resting on the paw that continued to hold the higher branch. Sometimes he would raise his head to look at me, his expression serene, his ears moving from the half-mast position to the upright, attentive pose, listening to my voice, for I always spoke to him when he decided to inspect me. But his scrutiny never lasted longer than a few seconds. Then he would give in to drowsiness. His head would nod jerkily, his eyelids slitted, until his chin was again resting on his paw and he was asleep.

By two-thirty that afternoon it became evident that the tom was not going to leave his tree for some time, probably not until dusk, so I decided to risk movement of my own, for I wanted to get closer to him. I intended to climb a tree in his vicinity, and would have to ford the creek to do so. By now I believed he would tolerate my presence in his immediate neighborhood, but I thought it would be better to wait for him to wake up before I put my theory to the test.

Fifteen minutes after I had decided this, the puma opened his eyes and looked at me. This time he didn't raise his head, and his ears remained relaxed. I got up, put on the backpack, and began to walk toward the water's edge, talking continuously and returning his gaze. Earlier I had marked a white-bark pine that grew on the west side of the creek no more than forty feet south of the tom's hemlock. These pines, *Pinus albicaulis*, are fairly crooked trees, the branches of which are widespread and offer themselves as perches from which vision is hardly obstructed. Trees of this variety that grow in open locations, or on rocky places, are usually stunted, almost shrubby, but those found in more sheltered areas reach a height of about forty feet. The one I was aiming for had attained its maximum growth and divided into two trunks at a place four feet from the forest floor.

The puma watched me intently, but otherwise did not change his stance as I strolled casually toward the creekbank, where I stopped and looked up at him. He turned his head and pricked his ears forward, watching me with that inscrutable gaze of the feline. He didn't move. I stepped into the icy water and felt it climb up my legs until it reached above the mukluks. The chill was intense, numbing. I kept going and the streambed started to rise again. The falling level did not, however, ease my discomfort. My footwear, socks, and pant legs were soaked; the cold seemed to be eating right into my bones, making me regret my rashness.

When I climbed out of the water I was no more than twenty feet away from the puma's hemlock, but I didn't waste time looking up at him as I sat on the ground to remove the mukluks and socks, draining the former and wringing out the latter, noting that the skin of my feet and legs had turned a light shade of ivory, mottled with bluish marks. Inside the pack I carried two pairs of heavy woollen socks, but I didn't intend to put them on until I had found a secure perch in the pine. Instead, I took out a towel and began to dry myself, rubbing vigorously to restore capillary circulation, looking up at the puma as I worked and chatted.

The tom had turned his head, lowering it so that he could watch me; two-thirds of the thirty-inch tail jerked from side to side, but he didn't appear disposed to move.

I shouldered the pack again and walked to the pine, carrying the dripping mukluks and socks, which I intended to hang from a branch. Climbing barefoot, my wet pant legs rolled up, I settled on a limb growing out of the northernmost of the pine's two trunks about twenty-five feet from the ground. I was now about forty-five feet from the puma.

Removing a few small obstructing branches, I had an excellent view of my wild neighbor, who continued as before, eyes fixed on me. We stared at each other for some moments before I busied myself with matters of personal comfort.

I opened the pack and took out dry clothing, setting a pair of pants and the two pairs of socks on a nearby branch before suspending the footwear and the wet hose from another tree limb. I stood on my perch, leaned against the tree trunk, removed my wet trousers, and climbed into the dry pair. Sitting once more, I again toweled my feet and legs until they started to tingle, and put two socks on each foot, and afterward donned an extra sweater underneath the parka. Thus accoutred, I felt infinitely more comfortable, ready to take the field glasses and focus them on the lion.

From my new vantage I could see every detail of the animal. He had returned to his old posture, head resting on his paw, tail hanging limp, eyes closed. He was dozing, but his ears were on duty, flicking this way and that continuously and pausing very briefly whenever they were aimed toward me. Without opening his eyes, he turned his head and snuggled it down on his forepaw, much as a human might do when settling more comfortably on a pillow.

In the face of the cat's astonishing tranquillity, I opened the backpack once more and took out the bag of trail rations. Eating, I was glad I had brought more food than would have been required for a twenty-four-hour journey. I had already been away from base for thirty-seven hours, and the amount of nuts, raisins, and oats remaining in the bag was barely enough to maintain me just on the edge of outright hunger if I managed to get back home by nightfall.

Having established to my satisfaction that the puma was willing to accept me in his immediate vicinity, I think I would by that time have ended my vigil and returned to base, had this not been my first contact with him. As it was, I was most anxious to prolong our time together in order to strengthen our relationship, to prove to the tom that I offered neither threat nor competition. In any event, I was determined to be present when he decided to leave the tree, no matter how long the wait. In addition to this, my mind was fairly bursting with unanswered questions: How long would he rest? Would he descend at dusk or earlier, or stay in the tree until after dark? If he descended while there was still enough light for me to see him, how would he react if I started to climb down my tree at the same time that he was descending his? These and other queries presented themselves as I watched and waited, doing my best to ignore the chill and to keep my limbs and backside from getting numb.

The thrill I had experienced when I first sighted the cat had long vanished, replaced by feelings of satisfaction. Even so, after more than three hours of watching him, I was becoming jaded with even that experience. He just wasn't *doing* anything!

Having eaten and made myself a little more comfortable on my hard seat, I began to note subtle signs that I had missed hitherto. A red squirrel scampered through the treetops, jumped into a young hemlock nearby, and chattered shrilly, disturbed as much by my presence as by the puma's. I had noted the little animal's arrival and was expecting it to voice its alarm when it saw us. For this reason I kept the field glasses fixed on the tom, wanting to see how he reacted to the squirrel's presence.

The cat's ears were obviously tuned in to the redback's movements and kept changing their direction as the arboreal rodent traveled. When the squirrel stopped, the puma's ears stopped also, fixed on target. So far he hadn't given any other signs of awareness, but when the squirrel began to chatter, the tom moved his head and wagged the tip of his tail for a few moments. Then he relaxed again, closing his eyes and allowing his ears to droop, his tail becoming still. He had evidently identified the intruder by sound, recognized at once that it was an animal too small to warrant the exertion of a swift chase through the trees, and dismissed its presence. I supposed that if it wasn't worth eating and didn't pose a threat, it would be a waste of time and energy for the puma to continue to track its progress.

The puma's reaction caused me to recall two similar disturbances that had occurred earlier, the first when a pair of ravens flew low over the trees and the second when a snowshoe hare passed underneath us on its way toward the creekbank. The puma behaved toward those visitors much as he had done toward the squirrel, alert to their passage and undoubtedly determining where and what they were mostly by hearing, though I noticed that his nostrils twitched as well, suggesting that he was also detecting their scent. As soon as he had evidently identified them, he ignored them. Thinking about this, I couldn't help realizing that he was treating me in much the same way, alert to my movements, sometimes twitching his tail, and occasionally lifting his head to study me more closely. This reinforced my earlier supposition: He had classified me as harmless, a being that was neither edible nor to be feared. Was it possible, I now wondered, for us to build up some sort of friendship, as Tom and I had done in London?

Musing in this way and looking more or less casually at the tom, I failed to notice that he had been bunching his body, preparing to rise, until he was in the act of doing so, but I did see the way in which his muscles flexed and corded along his left haunch and side just before he rose to all fours with absolute grace. For a few seconds he balanced easily on the narrow branches, then he turned to look at me before indulging in a luxurious stretch, yawning at the same time. His back arched acutely, his claws extruded, his great mouth gaped.

As he was standing broadside to me, I was given a profile view of his fangs and tongue. The great tusks were the color of old ivory; his tongue, pink as a boiled ham, arched at the back and stretched forward at the same time, curling smoothly at the tip. He ended the yawn suddenly, snapping his mouth shut, but he came out of the stretch slowly, relaxing his pose languorously and causing his muscles and skin to ripple as they returned to normal. He began to wash himself, licking a forepaw, passing it over his muzzle, licking it again, reaching backward to do an ear. Over and over he repeated these actions, systematically cleaning his head and face from side to side and from pate to chin. When finished, he pressed his chin down as far as it would go, resting its nobby tip on his upper chest before sticking out his elongated tongue and washing his front as high as he could reach. When he was through with this part of his toilet, he shook both forepaws one at a time, flicking them rapidly from side to side while holding them forward, elbows bent, wrists limp; afterward he twitched his ears, then sat upright on the lower branch, using his right forepaw to steady himself against the trunk of the hemlock before starting to lick his lower chest, belly, and genitals. This ended his grooming. He stretched and yawned again, but not as slowly as before; when that part of the ritual was over, he pivoted on his back feet, facing the tree, extruding his front claws and reaching upward, well above his head, to scratch vigorously at the bark, the powerful claws ripping off large pieces that fell to the ground. He reminded me of some huge domestic tabby working on an oversized scratching post!

Soon he would vacate the hemlock and glide away into the forest in quest of prey. There I could not follow him tonight, because the moon was barely in its first quarter, so I determined to wait for him at the foot of the pine.

I gathered my belongings, stuffing the wet garments into the backpack and struggling into the damp mukluks before shouldering the pack. The tom was still clawing at the tree when I started to climb down the pine, but I had only descended about ten feet when he stopped working his claws. The light was now failing rapidly, but I managed to distinguish his shadowy form as he abandoned work on the trunk and dropped to all fours on the branch. I kept moving downward and so I lost sight of him, but I heard him when he embraced the trunk and began sliding down it tail first—the racket he made was considerable until he reached a branch that had to be negotiated. As the noise stopped I looked up, seeing him only about ten feet up, a darkling shape silhouetted against the pallid sky. He stood with feet bunched together halfway along the last dead limb of the hemlock, clear of the foliage.

I hurried, wanting to be first on the ground, but the backpack caught on a small dead branch that broke with a loud crack. I was ready to swear, believing that the sudden noise would cause the tom to speed his descent. But the sound must have had the opposite effect on him; he remained poised on the branch and I was able to reach the forest floor ahead of him. But only just! I had hardly positioned myself beside the pine when I heard him move. He was descending headfirst, running down the almost vertical trunk and producing loud scratching sounds. When he was about eight feet from the ground, he kicked against the hemlock with his back feet, propelling himself upward and away, sailing through space with his front legs bent at the elbows, paws outstretched, his body slightly arched and his long tail streaming like a pennant. He hit the snow with his front paws while his hindquarters were still in the air, touched down with the back feet positioned almost between his slightly spread front legs, and then, with the merest pause, launched himself again. His ground leap covered 15 feet, $3^1/_2$ inches (by my later measurement), a beautiful, easy jump that lifted him about five feet from the ground to sail past me as a tawny blur. He had moved so quickly that I was unable to note his actions in more detail, even though he could not have been more than six or seven feet from me as he flashed past.

When he landed the second time, he appeared to check his forward momentum by stretching his front legs, opening them wider, so that his spread paws acted as brakes, but he made two more jumps, each shorter than the last, before he stopped forty feet away and turned his head to look at me, his ears erect, his tail lashing spasmodically from side to side.

We eyed each other for a few seconds. That is to say, *he* looked at *me*, for I could only see the outline of his body with any degree of clarity because of the falling dusk and the distance between us. But he appeared to be relaxed; and the fact that he stopped to examine me was encouraging, for it meant that he was certainly not afraid of me. Nor was he in any way aggressive. At no time during our brief period of close contact did I feel apprehensive. On the contrary! Without really pausing to think about the matter consciously, I felt at ease and confident that the puma and I had entered into a relationship of mutual trust. In any event, even if I had not been entirely convinced that such was the case, the tom's next actions would have banished lingering doubts.

He finished inspecting me, turned to face front, started forward at a walk, then stopped abruptly, sitting on his haunches with his back to me as he raised a hind leg, cocked it high over his shoulder, and scratched vigorously at his right ear. The sound—like a sharp tool rasping across dry leather—reached me clearly across the distance. When he finished, he twitched the ear several times and made as though to scratch again, but changed his mind. Without a backward glance he rose and moved forward, all in one fluid motion, disappearing like a wraith among the forest trees.

No wild animal ever turns its back on another being unless it is absolutely certain that no harm will come from the action. That one simple gesture was of greater significance than all the tom's previous actions. Without a doubt, he trusted me fully!

Tired, cold, and hungry, but elated to the point of euphoria, I began to retrace my way home. No amount of personal discomfort could possibly dampen my exhilaration, despite the fact that before I had covered the first mile of the journey, darkness overtook me and slowed me down to a snail's pace.

I do not mean to suggest that walking through a darkened wilderness is an unpleasant experience. It is not. But in the absence of the moon or some other form of lighting, progress is slow and orientation is sometimes difficult. When driven by hunger and cold, a journey of only a couple of miles can seem interminable and one must then harness one's full stock of patience and subdue the temptation to force the pace and become careless, risking injury or becoming lost in the darkness.

I actually enjoy traveling through the night forest. The wilderness becomes another world after sundown, a new environment to explore that offers greater challenge and a deeper understanding of natural events. Long before that night in the Selkirks I had learned to travel safely through the sylvan darkness, even to the point where it no longer required an effort of deliberate will to switch the eyes from daytime to low-light efficiency.

Within each human eye there is a depressed section in the middle of the retina that provides the keenest vision during conditions of light. This little behind-the-lens saucer is called the *fovea*; it always strives to pick up light energy, even at night, but its efficiency is greatly impaired in low-light conditions. The peripheral edges of the retina, on the other hand, have been designed to take advantage of even the weakest glow, and just as they are especially sensitive to movement during the day, they are also able to take over the basic functions of vision from the fovea during darkness.

Those not accustomed to traveling through the night wilderness may imagine that such an experience is fraught with difficulties and danger. In reality, though care should always be taken when traveling in wild places, the special mechanism of the eyes can serve remarkably well after sunset if an individual learns to take advantage of *all* the properties of vision.

Popular opinion to the contrary, man *can* see in darkness.

The whole retina—the membrane that lines the interior of the eye—is the only part of the visual system that is sensitive to light, because it contains an abundance of receptor cells called rods and cones, according to their shape. Within the retina of each eye are to be found about 125 million rods and some $6^1/_2$ million cones.

The greatest number of cones are found in the fovea and combine to produce frontal vision. But the more numerous rods are concentrated within the peripheries of the retina.

Individuals who constantly use artificial light during the hours of darkness never really give their eyes a chance to function at full capacity and are often not aware of the astonishing efficiency of these organs, which can actually perceive the light of a single candle across a distance of fourteen miles, once they have become well adapted to darkness. *

In addition, a normal human eye becomes up to one thousand times more sensitive to light energy after it has been

*Established by Dr. Selig Hecht et al. at Columbia University.

dark-conditioned for between twenty and sixty minutes, this sensitivity due largely to peripheral vision, which is even more important in poor light than in full daylight.

Years earlier I had learned to travel through the night wilderness by walking slowly, swinging my head rhythmically from left to right, scanning the environment with peripheral sight, even during full darkness. At the same time I alter the length of my stride, shortening it, and change the normal position of my foot, lifting it only an inch or two above the ground and causing the toes to point downward continuously, sliding each foot into contact with the forest floor. By feeling my way along the trail with my feet, I am able to keep my eyes scanning the way ahead and to the sides, from waist to head level.

There is a bonus, too! Because one is working blind, all the senses are sharpened. Hearing improves, as does the sense of smell; even the sensitivity of the exposed skin is heightened to the point where it is possible to detect the direction from which the wind is blowing, not only by direct contact with the skin, but through the hairs on arms, hands, face, and head. Lastly, by walking as I have described, one discovers that the toes and the ball of each foot are equipped with a sense of touch that functions even when one is wearing heavy-soled boots.

It took me three hours to cover the four miles between the puma's hemlock and base camp. And I was certainly glad to get home.

For almost a week after my first sighting of the puma, I remained close to base, waiting for the moon to ripen, keeping busy by plotting the information I had gathered to date and tracing on the map the range of the tom as accurately as was possible at this stage.

Most of the signs left by the male puma up to now—including the sighting and his inspection of me when I fell asleep while on watch—suggested that the cat's immediate home ground was located along both sides of French Creek, extending up the mountains on either side of the valley to an altitude that varied between three thousand and five thousand feet, depending on the rate of rise, the terrain, and the availability of prey. This area encompassed about fourteen square miles of wilderness where caribou, moose, goat, and deer as well as a variety of smaller game lived. In addition, the tom foraged, at least occasionally, on both sides of the Goldstream River, from its junction with French Creek eastward at least to the end of the more southern valley, a region of approximately thirteen square miles. The two territories combined to form a total range of between twenty and thirty square miles.

The second cougar, whose smaller tracks I had found and which I believed to be a female, occupied the area west of French Creek, extending its domain westward almost to the place where the power line crosses the river (my point of embarkation) and eastward at least as far as French Creek, perhaps beyond it, so that the range of both animals almost certainly overlapped to some extent. In addition, the female cat used the Goldstream River valley and the small mountain that lies between the two waterways and the Revelstoke highway, a tor attaining a height of 3,900 feet, but sloping gently and bisected by many game trails. The second puma's total range, I estimated, came close to the tom's, about twenty-five square miles.

When I had finished making all these calculations, I realized for the first time just how enormous a task I had set for myself and I concluded that I could not hope to study both animals to good effect over a section of mountain country that contained at least sixty square miles of cat range. For this reason I decided to concentrate about eighty percent of my energies on the tom, devoting the rest of my time to scouting the female's territory.

Beyond such calculations, I had also compiled a list of the animals that I knew beyond question lived in the region, dividing these into two categories: mammals and birds. Next I listed the trees, shrubs, and plants I had noted. When this task was finished, I found that I had identified 39 species of mammals, 101 species of birds (in season), 27 species of trees, 22 species of shrubs, and 39 species of flowers. Although I would have loved to obtain counts of the numbers of individual mammals that inhabited the region, this task was quite beyond my resources, but I did manage to determine that moose were the most numerous big-game animals and were followed numerically by caribou, mule deer, and goats; I had found the tracks of solitary elk in three different locations, but these were clearly isolated wanderers that might not remain in the area during winter.

After concluding these necessary chores, I spent time evaluating my knowledge of the puma, compiling two separate summations, one dealing with information obtained at first hand, the other confining itself to data contained in the literature or obtained during talks with other observers. This task was absorbing at first, especially that part of it dealing with my own observations, but toward the end I became frustrated to realize that beyond purely biological data, the sum total of knowledge dealing with the animal's personality, and especially with its behavior in the wild, was limited indeed. There is in the literature a plethora of statistical details, but these are confined almost exclusively to matters of physical biology such as size, general anatomy, gestation, coloration, dentition, and so on. These things are dealt with minutely, whereas information concerning the lion's habits is all too frequently based on hearsay; or else the isolated behavior of one lion is deemed to apply to all other members of the species.

My dislike of generalizations notwithstanding, I realized that I would also have to resort to describing the mountain lion according to the habits I observed in the tom and, if I was lucky enough to see her, those I observed in the female. This, and the knowledge I had gained by observing other pumas elsewhere, would help me to build a better picture of the puma, but I was only too well aware that I could not hope to offer a definitive study of this elusive animal. Indeed, such a task is beyond the ability of any one observer, for, despite the recent increase in sightings in North America, the peak of lion populations passed into history some two centuries ago. Since then, region by region, man has claimed more and more of the big cat's range while making almost incessant war upon it, and although modern conservation practices, combined with the creation of game preserves, appear to have ensured its continued survival, I believe that today's cougars, having been mercilessly persecuted, have virtually altered their personalities and have become more furtive than their progenitors of two hundred years ago, which, according to many early accounts, were much more likely to move about openly in the presence of man.

All animals that are constantly hunted become wary. This is a fundamental law of survival. But when they have been hunted unnaturally (by man, using dogs and the weapons of modern technology) and when their habitats are invaded by machines and cleared for farming, or drastically altered by logging and mining so that shelter and food sources are decimated, those animals that manage to hang on do so by becoming supercautious. If such threats continue for a long time, the offspring of the survivors learn from their parents and grow up with the knowledge that man is an enemy to be carefully avoided.

If only because of altered ranges, mountain lions today differ physically from their precursors. In many parts of their territory they are smaller; they have also altered their diet because in some regions they cannot now depend on adequate supplies of large game animals; and in those areas in which they have not been exterminated, they have adopted new survival patterns, living in habitats where their kind would not have been found two centuries ago, such as within deep, coniferous forests where only a few large prey animals are to be found now and then, and where food consists of small animals, such as hares, squirrels, porcupines, and some game birds. In some instances, confused young pumas, faced by hunger and attracted by the scent of domestic stock, inadvertently wander into the thick of a populated area, where they are immediately destroyed. Such an incident took place in August 1981, when a two-year-old, 75-pound tom, undoubtedly confused by traffic and the cacophony of human habitation, entered a brewery in New Westminster, British Columbia. This animal was immediately shot by a government "wildlife control" officer. It could have been tranquilized and returned to a suitable habitat, but it was clearly easier to kill it.

Before the arrival of Europeans in North America, the Indians hunted the puma, of course, but the odds were always in balance. It was then a case of human reason against cat cunning and strength, a battle of wits rather than a one-sided, ruthless war of extermination. But once the whites had obtained a firm foothold on the continent, the forest began to shrink under the assault of pioneer axes and saws; guns killed at long distances, and the odds shifted in favor of man. Still, the puma held on throughout its range. By stealth and great wit it continued to elude its human pursuers more often than not, loping away silently, then taking shelter high in a tree, whence, camouflaged by its monotone coat and as still as a statue, it must frequently have watched as the hunters ran past its refuge. But then came the hounds of Europe—the noses of man. They led the hunters with their guns to the trees in which the pumas were sheltered, discovering for the newcomers the weakness of the lion, which, like all cats, has never been a long-distance runner and soon seeks a safe hiding place when disturbed.

Historians have made much of the fact that the Indians hunted with dogs; and so they did, but their animals, wolf-dog crosses, did not bark during a chase, a lupine characteristic still to be found in today's huskies and malamutes. Such dogs can be used for close work, such as bringing down an animal or keeping it at bay while the hunter delivers the killing blow, but in a chase they soon outdistance the hunters, and since they remain silent while they run, their masters cannot locate them and the quarry by sound. But the Europeans had been breeding specialized hunting dogs for centuries, and their hounds, selected for the deep, baying bark that they utter compulsively during a chase, proved ideal for puma hunting. Even so, by further careful selection, special lion hounds were developed on this continent. As a result, the odds have now been stacked against the cougar.

From my own experience as well as from the literature, I know that, like many large mammals, the puma does a lot of traveling and may take two or three weeks to complete one circuit of its range, although there is no assurance that it will stick to a timetable. Too many variables are involved. If a cat has pulled down a large animal, such as a moose, it may stay in the area of the kill for several days, eating, then raking leaves and debris over the kill before finding a nearby shelter in which to sleep until it is again hungry. If game is plentiful, it may not need to prowl far and may feed only once off the same kill, leaving what it does not consume for the scavengers of the wilderness. If game is scarce, it may roam for miles, or even leave the country altogether, while a female with kittens will remain near her den until her young are old enough to travel, usually when they are about two months old and begin to go on night hunts with the mother, staying in the background, but learning by watching her.

Such variables can influence a puma's movements and cause would-be observers no end of frustration, but, as if to make up for its irregular habits, the big cat almost always travels along the same trails, which thread mazelike through the wilderness and often cross each other at particular locations. This practice is not, however, dictated by an inordinate love of routine, but rather because all animals prefer to travel along the easiest paths for the same reasons that prompt members of the human race to seek the most effortless route for themselves. This being so, wilderness game tracks are used year after year by predator and prey alike when they travel to and from feeding or resting areas, their movements timed according to their needs and whims as well as by seasonal changes that alter the availability of herbivorous foods and cause the prey species to forage elsewhere. In turn, this also affects the predators, which may forsake a trail until the next growing season causes the game to return.

For these reasons, some trails are used heavily during the summer, others are used in winter, and still others are traveled the year round, depending on location and the country through which they wind. But this is not to say that game trails are always crowded with a constant procession of animals; as a rule, both the prey and the predatory species use the pathways cautiously—sometimes, if they are herd animals, trotting as a group but in single file and well spaced, but always ready to dart into the forest at the slightest hint of danger. Hunting animals, such as pumas and wolves, use the trails in much the same way, except that the cat is alone and that both species are always alert for the sound made by prey animals.

When a man walks these meandering paths, he is unlikely to sight any living thing along the route, because the noise he makes acts as a warning; but if he sits quietly, just off the trail, he is quite likely to see a variety of animals within the space of an hour or two.

The puma's pathways, therefore, are not the cat's exclusive thoroughfares, but are merely parts of a vast network of such tracks that generations of wild animals have kept open in forest and jungle everywhere in the world. The lion, however, likes to mark its trails with its scratch hills, especially where one intersects another. On such crossings the cat may stop repeatedly to urinate or defecate, in time leaving a dozen or more of its distinctive markers. This habit has caused biologists to believe that the puma marks its territory against others of its kind, a theory that I found interesting, but one that was seemingly arrived at without any evidence other than the hills themselves. In the absence of proof, I had accepted this premise with caution, but later, during my Selkirk research, I was given cause to reject the notion.

By the time all the paperwork was done and I had finished charting the range of the two pumas, the moon had grown into its second quarter and the period of cloudy weather that had persisted for several days showed signs of lifting. Now I spent one more day preparing myself for the arduous field-work that was about to commence, inspecting and sorting the equipment that would allow me to remain self-sufficient while I devoted all my energies to the task of tracking and following the cougar. I had already constructed the three emergency shelters at the extreme ends of the country that I was going to study, but because I could not count on always being within reach of these, I had to gather together materials that would allow me to bed down on short notice anywhere within the puma's range. For this purpose I had brought a piece of heavy plastic 12 feet long by 10 feet wide that was fitted with brass grommets to which were attached lengths of rope 24 inches long, for ties. Stretched and secured over a framework of saplings and covered by evergreen boughs and snow, the plastic would furnish a water- and draft-proof shelter that could be erected in less than half an hour.

To lessen weight, I kept my hardware down to a minimum, discarding the bow saw in favor of a two-foot length of bone saw, an expensive bit of modern technology that looks like a length of stainless steel wire fitted with little burred teeth; handles of wood are fitted at each end, so that it can be sawed back and forth around firewood logs up to eight inches thick. This gadget is not quite as easy to use as the bow saw, but it can be carried in a pocket. In addition to this, I would take a two-and-a-half-pound axe, fitted with a leather edge guard, and my heavy bush knife.

The wire bone saw, the axe, and the knife were all the tools I planned to take, except for the jackknife looped to my belt, without which I never travel. I also would carry two pocket hand warmers with a supply of solid fuel for use within the sleeping bag should my feet get cold, for chilled extremities will quickly cause the entire body to suffer; one dozen ordinary six-inch candles; a nest of camping utensils; my compass; fifty feet of quarter-inch nylon rope—for emergency descents or ascents in tough country—and two twenty-five-yard lengths of manila clothesline.

I was going to take enough food to last five days (fifteen pounds), so I packaged this in plastic bags, premixing enough flour, baking powder, salt, and powdered milk to make up daily supplies of bannock bread. The rest of the rations consisted of rice, dried beans, dried mixed fruit, roasted soybeans, rolled oats, whole-wheat kernels (for chewing on the trail), my staple mixture of oatmeal, nuts, and raisins, half a dozen packets of dried soup mixes, a dozen OXO cubes, tea, coffee, and one pound of powdered skim milk.

Added to this was my sleeping bag, one five-point Hudson's Bay blanket, and spare clothing, including four pairs of heavy wool socks and two pairs of cotton longjohns. Last came my snowshoes and a packet of rawhide for emergency repairs, two pairs of heavy leather mittens with wool liners, an extra pair of light canvas mukluks with felt liners, and a spare hat—a leather and synthetic-fur affair with ear flaps.

All this equipment was contained in the backpack and in a smaller belt haversack. The whole must have weighed about forty pounds, but I consoled myself with the knowledge that this total would be reduced by about three pounds daily, as I consumed the food. By the same token I knew that five days' worth of supplies could, in an emergency, allow me to stay in the field for seven days, perhaps even eight, if I was prepared to accept acute hunger.

By the time everything was ready on the eighth day after I had sighted the puma, it was already dark, so I had a hearty supper and turned in early, setting the alarm for 5:00 A.M.

8

DURING THE FIVE WEEKS AFTER I HAD SIGHTED the puma for the first time, I was lucky enough to see him and relate with him on ten other occasions, and I found five of his kills. By the end of this period, during which I had kept continuously on the move, I knew a great deal about the Ghost's personal habits and about the geography of his range; but, of greater importance, I had become convinced that he had accepted my presence in his domain. This made me feel that I had already succeeded in my quest; I had managed to enter the territory of an adult, fully wild mountain lion and been able to gain his trust. If my quest now yielded no other significant results, this fact alone made my endeavor worthwhile.

Of course, I wanted more! But as I thought about the way in which our relationship had progressed, I felt intense satisfaction. At first the Ghost had been drawn to me by curiosity, which he demonstrated when he inspected me while I was asleep near his old kill; later, beginning with the occasion when he had allowed me to see him and to approach his tree shelter, he exhibited cautious tolerance. More recently his behavior had led me to believe that he took satisfaction from my presence, a conclusion I reached after he stopped several times on the trail while walking ahead of me and turned around to look back, resuming his journey when I got within fifteen or twenty yards of him.

When the weather turned colder and the snow came to stay, my task became easier because his tracks were readily visible and because I could now easily determine whether they were recent or old. I began to see him fairly regularly.

By this time, aided by the snow, I had learned a great deal about his traveling habits and was able to map his preferred trails, discovering that when he was actively hunting he systematically patrolled both sides of French Creek along regular pathways, always walking against the wind and often making short forays up the flanks of the mountains, selecting the locations that offered gradual slopes and seldom climbing above the one-thousand-foot level. These side trips puzzled me at first, for they were undertaken in territories where the tracks of moose, caribou, and deer did not exist, but later I began to find places where he had killed snowshoe hares, grouse, and white-tailed ptarmigan, each of which he had carefully stalked, holding himself low to the ground, even dragging his tail in the snow, and whenever possible hiding behind bushes, rock outcrops, and the boles of large trees. Invariably, when he reached cover that positioned him between eight and ten feet of grouse or ptarmigan, he charged. His success rate appeared to be excellent, for out of seven such rushes that I found, he had only missed one bird, the story of each strike being clearly imprinted in the snow. Hares he dealt with in a somewhat different fashion. He stalked them first, then charged up to thirty feet, running them down after essaying two or three great leaps that brought him close to the quarry. Snowshoe hares are fast, agile animals that dodge frequently when closely pursued, but they almost invariably run in circles, being reluctant to leave their familiar home ranges. It was evident that the Ghost knew the habits of the hares. His fast, short rushes were always clearly marked in the snow and showed how he either matched the dodges of the prey with counter-moves of his own, or cut across country and intercepted it.

My own travel patterns had become routine by now. When there was no moon, or when it was obscured by heavy clouds, I followed the puma's spoor in daylight. When there was a moon to guide me, I tracked at night, usually from sunset to midnight, at which time I rested at one of the prepared shelters or, wrapped in the sleeping bag, dozed on top of a pile of evergreen boughs until the cold forced me to move, when I would continue tracking and observing until dawn and then return to base or find shelter in a lean-to.

While returning to the shack one morning, after tracking him without success the night before, I sighted the Ghost. It was full daylight and I was north of the Goldstream valley, descending a game trail that threaded through an area of rocky, sloping land dressed by sparsely growing lodgepoles and a few deciduous trees.

The evening before, as the sun was disappearing behind the peaks, I had crossed French Creek at a point near the valley of the shack and almost immediately noticed the puma's tracks. They were leading south, toward the river. I followed and discovered that for the first time since the snow had fallen, the Ghost was hunting along the north side of the Goldstream valley. At eleven o'clock that night I lost his trail when the moon, now in its first quarter, was obscured by clouds. I was near the lean-to I had built at the end of the river valley, so I made my way there and went to bed after having some food and starting a fire. Before dawn I set out again, retracing my route to the place where I had lost the spoor last night. From here I was led upslope and into the rocky area, which extended for about one mile. The game trail meandered through this locale, twisting and turning around obstacles so tortuously that for every one hundred yards, I estimated I only covered about a quarter of that distance in a straight line. As the sun, obscured by the clouds, was nearing the eastern peaks, I emerged from the narrow trail into a small natural clearing at the western end of which a landslide had occurred in the distant past. This formed an abrupt scarp of granite at the foot of which were piled the boulders and scree that had fallen away from the rock face. These were now covered by snow, but given the fact that seedling evergreens were growing there in scattered locations, it was evident that the slide was old. About thirty paces away from the rocks I saw the remains of a deer that had been carelessly covered by snow mixed with debris from the forest floor to a depth insufficient to conceal one back leg elevated by rigor mortis. I was about seventy-five yards away, standing at the edge of the trees.

About to approach the remains, I hesitated, then used the field glasses to scan the surroundings. I really wasn't expecting to see the Ghost, but when I focused on the rock scarp, a shallow recess in its center attracted my notice. Lying within this shelter, the tom was watching me, his body fully outstretched, his neck and head raised; his tail was lashing spasmodically.

Being so close to this kill made me cautious. I had already noted the marks of the death struggle as well as the tracks left by other deer as they fled the scene, the distance between each set of prints showing that the escaping animals had been leaping. Two clear sets of hoofmarks told me that the puma had attacked three mule deer as they were browsing at the edge of the forest. Other marks in the snow gave me to understand that the Ghost had first detoured through the trees, approaching from the west and upslope, to position himself behind and downwind of the prey. Three great leaps had brought him into contact with one of the deer, which had been hit violently and had rolled some distance before coming to a stop at the foot of a young lodgepole. Here the tom had seized the quarry, which was probably already dead, and dragged it about fifty yards to the middle of the small clearing.

Why, I wondered, had he elected to feed in the open? Then, as I used the field glasses to inspect the area, and before I was aware that he was watching me at the same time, I realized that from the place where the remains lay, the puma had a clear view of three trail mouths that emptied into the open space. Eating, the lion could keep each track under observation, while the heavy forest and brush that encircled the tiny dale offered the kind of obstacles that would be certain to produce sound, no matter what manner of animal sought to travel through it. My wild friend was a *thinker*, no question about it! He had employed strategy during the hunt and again when he selected the place to take his meal.

Afterward, when I saw him within the natural rock overhang across the clearing, my respect for his intellect increased further. He had selected a shelter that overlooked the open area, the kill, and the trails, and at the same time forbade any but a frontal approach. Years earlier, during my time as a soldier in World War II, we would have called such a position a "killing ground," a place carefully selected by occupying troops because it prevented encirclement and would force enemy soldiers to cross an open space and so place themselves at a disadvantage.

Musing in this way, I remained still for a time, watching the Ghost, talking to him softly, and noting that he was distinctly ill at ease. He was staring at me fixedly, his tail continuing to lash, his ears flattened backward. He wasn't snarling, but his mouth was open and his lips were wrinkled, revealing the yellow fangs. He didn't seem to be disposed to rise to his feet, but the way in which he occasionally moved his gaze to the kill, then swiveled it back to me, told me as clearly as words would have done that he would not allow me to disturb his food.

To ease his tension I started walking slowly downslope, in the direction of the river valley, speaking continuously and raising my voice slightly as I moved closer to the southern edge of the little clearing, staying in plain sight. When I had detoured around and away from the carcass and was about fifty paces from the Ghost's shelter, I stopped beside a medium-sized boulder that would serve as a seat. Much as I had done on the occasion of our first meeting, I turned my back on him while I brushed the snow off the granite, an action that required a considerable effort of will and was accompanied by a feeling of intense unease.

I had deliberately avoided the tom's eyes as I walked, but when I turned around, sitting down at the same time, I looked at him immediately, my heart pumping somewhat faster than normal. I need not have worried. The Ghost was lying in the same position, but his attitude had now changed. His head was down, chin resting on crossed front paws. He was watching, but his gaze was softer than before; his tail had stopped moving and his ears were at half-mast. He had understood my intent and was no longer apprehensive about his kill. The light had now increased. I noted that it was 7:16 A.M.

I composed myself on the rock seat, remaining relaxed, talking to him intermittently while he dozed, his long, tawny body perfectly relaxed, his breathing deep, causing his uppermost side and flank to rise and fall rhythmically. Occasionally he gave me a slit-eyed glance, then allowed his lids to close.

After two hours, becoming stiff and chill from inactivity, I got up slowly. This action instantly attracted my friend's notice, but when he saw that I was walking into the forest, he relaxed anew. I moved purposefully, not trying to hide my whereabouts as I skirted the rock scarp and climbed up-slope for some distance, seeking to restore suppleness and circulation. Twenty minutes later I returned to find that the Ghost had changed position and was now lying partly on his back, his hindquarters flat, his hind legs sticking up in the air, but his torso angled so that he could still watch me. I resumed my seat. The Ghost yawned, rolled entirely onto his back, and let the top of his head rest on the rock floor of his shelter. He could still see me, even if now from an upside-down position.

Apart from two gray jays, half a dozen ravens, a number of chickadees, and one red squirrel, nothing much was moving in our vicinity, which remained quiet for a considerable time. The birds were feeding off the deer carcass. The jays and ravens would land, peck lustily until they freed a piece of meat, then fly away to either hide the food in the forest or to perch on a branch and eat it. The tiny chickadees would sit on the frozen kill and peck at it until they had eaten several small morsels, after which they left, no doubt to search for dormant insects in nearby trees.

The Ghost knew perfectly well that his food was being raided, but he accepted the minor depredations with equanimity, not bothering to waste his energies charging after the birds. Where a younger predator might have sought to chase away the robbers, the tom, though he watched them from time to time, undoubtedly knew from experience that it was useless to pursue the birds, which, in such an event, would fly off, perch on a nearby branch, and merely wait until the coast was clear before descending again on the food.

Ravens are the boldest and most stubborn panhandlers to be found in the wilderness. The gray jays are almost as daring as the ravens, but whereas they will wait until a predator has left the kill before sampling the food, the big northern ravens will land while the rightful owner is still eating, whether this be puma, wolf, wolverine, bear, or anything else on four legs. I have seen ravens stealing food right from under the noses of a pack of timber wolves. While the older wolves ignore the pestiferous airborne black thieves, the inexperienced cubs waste a great deal of energy and work themselves into a frenzy trying in vain to safeguard their rations. Once, in the Yukon Territory, I watched as nine large ravens harassed a young wolverine into such a state of rage that the unfortunate beast appeared in danger of apoplexy. Every time the angry animal snapped at a raven, its teeth clicked together emptily, and while it was watching its intended target rise into the air, another bird would dart in and steal as much as it could. Growling and charging first one bird, then another, the wolverine was kept on the go for almost twenty minutes while the ravens made heavy inroads on the hunter's meager food, a snowshoe hare that had been less than half-eaten when the birds arrived. In the end, the outraged animal sprayed musk over the remains, now mostly skin and bone, and fled the scene, growling angrily and occasionally swatting at harmless bushes that lay in its path. Before he was more than ten feet from the remnants, the biggest raven in the flock sped down and grabbed the skimpy leftovers, despite the musk, and took off into the forest, pursued by its vociferous comrades.

It was almost four o'clock in the afternoon when the Ghost bestirred himself; by then, the birds had taken their fill from the remains. The puma rose to all fours, stretching, yawning, and generally repeating the routines I had observed when he was sheltering in the hemlock, right down to a thorough wash, though this time he sat to do his underparts. Afterward, keeping an eye on me, he loped down the snow-hidden scree to feed off the carcass.

I had been watching him for more than ten hours, with only three short breaks for exercise. During this time I had eaten half a pound of trail rations and had slaked my thirst with snow. Now, watching the big cat enjoying his meal, I felt ravenously hungry, and this, on top of fatigue, stiffness, and chill, made me decide to abandon the watch and return to base. When I left, the tom hardly spared me a glance.

Arriving at my small home just as darkness was about to settle over the wilderness, I found that a wolverine had tried to break in, the marks of his teeth and claws evident in several places on the wall boards near the ground. But my visitor had not tried to broach the shutters or the door, guarded as they were by the bristling nail points. Pleased that my defenses had been tried and not found wanting, I reentered the shack, lit the fire, cooked supper, and set the alarm to ring at 2:00 A.M. so that I could put fresh wood in the heater.

After a long rest, I went out again an hour before sundown, knowing that the moon was in its second quarter. But after five hours of searching I failed to discover recent signs of the Ghost. The trouble was that by now the trails were so criss-crossed with tracks of every kind, including my own, that it was impossible to separate fresh spoor from old. I returned to base and went to bed.

The next morning I was up before dawn and on the trail at first light. Since I didn't know where to start looking, I decided to do a random search along the trails that led to the river. I intended to try to find recent prints by walking through the forest, off the trails, where the snow was relatively un-disturbed; then, if I crossed the puma's sign, I would try to find the shelter in which he had settled down for the day. If I managed to locate him at rest, it would be safe to conclude that he had hunted successfully and would spend the hours of daylight taking his ease; should this be the case, I could return to base and relax there until late afternoon before returning to continue my observations during evening and night, when pumas are most active.

An hour after setting out on this quest, I reached the sandbars near the end of French Creek. I had found no fresh pugmarks, but had noted an abundance of recent tracks made by a variety of other animals, including moose and caribou. Wanting to get a better view of the river valley, I ascended the slopes, selecting a game trail that had at one time also been used by my own species. This runs from French Creek in a westerly direction, then turns sharply north and follows McCullock Creek to its source atop the next mountain, but the trees and shrubs concealed the valley from my view. Nevertheless, before I had traveled more than a quarter of a mile, I found very recent puma prints just off the trail. Back-tracking these, I realized that the Ghost had been using the beaten pathway to head westward, and that my failure to notice his tracks resulted from the hardness of the snow and from the many other imprints made on it.

Where the cat had left the trail, his spoor aimed upslope. I followed, walking slowly and as quietly as possible, because I felt certain that he had recently passed this way. Five minutes later, approaching a more open area that overlooked McCullock Creek, I saw the Ghost. He was about three hundred yards away, downslope from me and in the act of stalking! The breeze was westerly, blowing toward me. As usual, he was advancing into the wind, so that he could scent any prey in the vicinity while the animals he was looking for would not detect his own odor. By the same token, neither the Ghost nor whatever he was now hunting would be likely to detect me by scent.

Realizing that the lion must have been hunting all night and was probably very hungry, I moved with extreme caution, not wishing to distract him or the quarry he was stalking. I hid behind the trees and shrubs until I neared a point where I would be able to look down on the creek and on that narrow, sloping valley through which it ran. Moments later I saw a solitary female caribou pawing at the snow to uncover patches of cladonia, the so-called caribou moss that is in fact a lichen. The animal looked emaciated when I focused the glasses on her. The shrunken stomach suggested that she had not been feeding well of late, though why this might be so was difficult to understand, for there was abundant feed in this region.

The cow was standing broadside to me and looking north,

chewing slowly. The puma was advancing in total silence, pressing his body as low to the ground as possible and always keeping trees or shrubs between himself and the quarry. He was above the caribou, about thirty yards higher than she was. In front of the cat and in line with the cow was a rock outcrop, the top of which was some twenty feet above the floor of the narrow valley. The cow was more than forty feet away from the snow-covered granite upthrust; the puma was half that distance from the rock; that he was making directly for it was certain. I surmised that if he managed to reach the height undetected, he would leap at the prey from there.

The caribou moved slightly, turning so that I could see where one of her meager antlers had been broken off almost at the place where it joined to the frontal bone of the skull; the stub was ragged and the remaining antler was sparse and lacked the distinctive, forward-jutting "plow blade" characteristic of the bulls of the species. Indeed, though distinguished by being the only members of the deer family whose females grow antlers, those worn by caribou cows lack the spectacular spread and shape of their consorts'. Not all cows bear antlers, perhaps because some of the weaker or less well-nourished animals cannot spare the physical resources needed to grow them while they are nursing their calves. Those cows that do have them—and they are in the majority—do not shed them in the autumn and winter, as do the bucks, but carry them until April or May, losing them at about the time that they bear their calves.

The most notable difference between mountain caribou and their close relatives, found in the arctic, is size. The mountain species is larger, some bucks weighing as much as 600 pounds, although the more usual weight in British Columbia is 350 to 450 pounds, cows being about twenty-five percent smaller. Examining the animal that was now the puma's target, I judged it to be considerably thinner than most. Through the glasses and even across a distance of about 350 yards, I could count her ribs; and her coat, which should have been thick and glossy by this time of year, was clearly in poor condition.

I didn't spend much time observing the caribou, because when I altered the angle of the glasses to look at the tom, he was already bellying up the granite crag, his progress slow, silent, and so calculated that I could almost visualize the workings of his mind. He moved one limb at a time, positioned it, then moved another until he had advanced about half his own length; then he wriggled his hindquarters, settling himself on the new place firmly but gently, so as to prevent any sudden noise escaping from some limb or twig lying concealed by the snow, which could snap as he began his next move. I too had learned to crawl in this way, and although my own actions were somewhat different from his because of my less efficient anatomy, I knew that it was important to carefully flatten the ground and the debris that covered it before making another move.

After three or four minutes of infinitely patient crawling, the Ghost stopped. His eyes and the top of his head must have been a few inches above the highest point of the outcrop, but the cow failed to detect his presence. Keeping the glasses focused on him, I saw how he positioned his haunches for the spring, his muscles rippling smoothly under the golden hide, his shoulder blades moving visibly, his neck tense. The tail lashed silently for some seconds, then was stilled and held rigid. He leaped. He sailed through the air, his body an arc, his tail a pendulum; his massive front paws were held directly in front of his chest by his stiffly outstretched legs, but the toes were not spread and the talons were sheathed.

From his elevated perch, the puma covered thirty-two feet in his first enormous leap. His forepaws touched the ground as the caribou realized her danger. But before she could do more than raise her head, the Ghost sprang again, closing the intervening distance and hitting the cow on the right shoulder with both forepaws. The blow was devastating. The caribou's long neck was snapped violently to the right, her vertebrae cracking so loudly that I heard it break. In a trice her inert body was hurled toward the creek, rolling over and over down the shallow grade while the puma leaped behind it. The cow had been killed by that first mighty, whiplashing blow. Even so, the moment her body came to rest, the tom landed on it. In swift sequence, as his hind feet came to rest on the animal's back, he reached out with his left forepaw and grabbed the cow's nose, pulling it violently toward himself as he clutched the quarry's neck with his right paw and at the same time sank his fangs into the back of her neck. When the tom realized that resistance had ended, he let go. The caribou's nose was deeply gashed from the razor claws; the neck bled profusely from the fang bites.

Standing beside his prey, the Ghost sniffed her carcass slowly, his muzzle actually touching the animal's hide. When he reached the neck, he tentatively licked at the blood that oozed from the wounds, then reached forward with his right paw and, claws fully extruded, ripped her paunch open with one swift, sure slash. That done, he squatted beside the carcass and began eating the warm entrails, liver, and lungs. For half an hour he fed, taking his meal mostly from within the animal. At first I could see steam rising from the body opening, but as the meat chilled, only the cat's exhaled breath was visible.

When he had taken all he wanted from the cavity, the puma withdrew his head, most of which had been buried inside the carcass. He turned, allowing me to see his face. It was a mask of horror! Blood smeared it so thickly that it dropped off in places; pieces of intestine, some discharging partly digested vegetation, adhered to his fur in several places.

When he licked his lips and I was able to see inside his mouth, all his teeth were red, and his normally pink tongue was scarlet.

He returned to his meal, holding down the carcass with both front paws while he ripped at the haunch that lay uppermost, biting off pieces of skin and fur and spitting them out. Soon he had uncovered an area about the size of a dinner plate. A thin cloud of vapor escaped from the hot meat. The tom bit into the haunch, closed his jaws, and pulled mightily, at the same time keeping his front legs stiff so as to hold down the hindquarters. When the mouthful was freed, it made a sound similar to the sudden tearing of a piece of cloth.

After he was fully sated, the Ghost sat up beside the carcass and began to wash the gore from himself. Methodically, fastidiously, he spent forty-two minutes cleaning his face, head, chest, and paws; then, looking presentable again, he took more meat from the exposed haunch, but interrupted himself in mid-chew in order to turn his head to stare in the direction of the Goldstream River, out of sight to both of us, but only about a quarter of a mile away. Flattening his ears, he leaped to his feet and stood beside the carcass to face the creek mouth. He had clearly been alerted by something beyond my senses, but whatever it was did not appear to have the power to make him run away and leave his kill. He stood like a statue carved out of amber, except for the lashing tail. A moment or two later he opened his mouth and curled his lips; he was growling softly, a sound that only just reached me across the distance.

Now I too heard noise, a series of crackling sounds coming from the forest on the west bank of McCullock Creek, but near the narrow valley. The disturbance advertised the imminent arrival of more than one large animal. Counting time instead of allowing myself to be distracted by looking at my watch, I waited expectantly, my gaze alternating between the immobile puma and the far bank of the creek. When I reached 139, which told me about two minutes had passed, I saw the bushes move at a place about seventy-five yards south of where the puma and his kill were located. Instants later, the wolf pack came into view. It was White Spot, the big lead wolf I had seen on the bank of the Goldstream while I had been ferrying the last load of supplies to the valley of the shack. But now his pack consisted of six wolves instead of seven.

Tails erect, heads held high, the six wild dogs trotted toward the Ghost, with White Spot in the lead. The tom stood his ground, his growls deep and loud and malevolent, his mouth agape, the great fangs ready for action. I didn't really believe that the wolves would be foolhardy enough to attack the big puma, but it seemed as though they were about to do just that. Then, as the Ghost bunched his body preparatory for a charge, White Spot veered to the right, followed by his bitch. The four other pack members swung left.

Moments later, without signals or any other kind of instruction from the leader that I was able to detect, the wolves formed a loose circle around the puma. They trotted slowly around the kill, heads turned toward the cat, but keeping about thirty feet away. For a short while the scene was reminiscent of one of those Hollywood westerns where a besieged wagon train is experiencing an Indian attack. But the Ghost put an end to the encirclement when he made a short, furious charge that almost caught one of the wolves off guard. Missing it by inches, the cat immediately returned to the kill and White Spot broke away, retreating about twenty feet and lying down almost casually, but facing the cat. The other wolves followed their leader's example. Now the Ghost and his kill were in the middle of the pack of silent wolves, all of which stared at the dead caribou, mouths agape and salivating, ears pricked forward, their faces reflecting keenness, but devoid of savage overtones. They reminded me of my team of sled dogs just before it was time for their evening meal, each animal expectant, but wearing the same cheerful expression.

The puma also relaxed. He sat beside the carcass, and although his ears were still pasted against his head and his tail lashed constantly, he did not seem greatly perturbed. I timed the action (or *in*action) with my watch. For eleven minutes the animals remained passive. After the first three minutes had elapsed, the Ghost began washing himself again, though he kept a sharp eye on the competition. The wolves meanwhile yawned and some of them wagged their tails; two of them whined eagerly now and then. The entire pack watched the puma intently.

Seconds short of the eleventh minute, White Spot got to his feet and made as though to advance on the puma, who immediately rose also; the wolf walked a little closer to the cat before stopping. Meanwhile, his bitch had risen as well, but on the far side of the tom. Suddenly she charged, running in swiftly only to be met by a growling, spitting bundle of fury. The bitch dodged the puma's counter-charge and sped away, and at this moment White Spot initiated an attack of his own. He, too, was met by the furious lion and was forced to dodge and retreat, but another wolf took his place. This kind of puma-baiting went on for several more minutes. It was clear that the pack was trying to stampede the tom, to cause him to give up his prize and to seek shelter in the forest. But the Ghost was made of sterner stuff. He became louder and angrier with every charge of his enemies; his tail lashed more swiftly, and his ears were finally pressed back so tightly that they were almost invisible, making his head look as smoothly rounded as a seal's.

By the time that all six wolves had each made several charges, the Ghost took countermeasures. Instead of trying to reach a wolf he had selected as a target, he stopped suddenly, twisted agilely, and almost caught the bitch, who had been about to attack while the puma's back was turned. The furious cat came so close to the bitch that she dodged frantically, stumbled, yelped once, and ran straight for the shelter of the forest.

This ended the siege. White Spot turned away, but he walked sedately, as if to run would be beneath his dignity. The other four wolves were quick to follow their leader.

The Ghost, having routed the enemy, astonished me by turning to the carcass and biting into it. His fury had dissipated as though by magic.

As for me, I was covered in sweat and feeling decidedly limp. I had experienced so many moments of excitement since first sighting the caribou that I was emotionally spent—so much so, in fact, that I decided to return to base and rest up for at least a day. But first I wanted to see what the Ghost would do when he learned that I was in his vicinity, reasoning that if anything was going to strain our relationship, my appearance so soon after he had made his kill and immediately following his altercation with the wolves would surely do it. I was somewhat reluctant to put the matter to the test, yet I felt impelled to do so, because if he showed forebearance under such circumstances, any lingering doubts I might have about his intentions toward me would vanish.

Nevertheless, I did not intend to approach him too closely. My presence on this occasion would surely come as a surprise, for he had first been so preoccupied with hunting that all his attention had been riveted on the business at hand. Afterward he was almost equally occupied as he fed on the fresh meat, and later, when the wolves arrived, he was forced to concentrate on the challenge that they offered, making this the first time that I had been able to observe him without his being aware that I was doing so.

I started to walk to the rock outcrop, where I was going to stand in full view; I moved casually, not making much noise, but not trying to conceal my movements. I had taken only twenty-six paces when the Ghost sprang to his feet and whipped around to face the rock outcrop, above which my head and shoulders were visible. His ears were flattened, his tail was lashing, his mien showing clear annoyance. I started talking to him as I walked more slowly before finally stopping on top of the rock, where I squatted, lowering my height and thus reducing my bulk, which minimized any threat that he might see in my arrival. Animals are more impressed by height than by girth when assessing the potential danger offered by an intruder, and although the large predators do attack and kill animals taller than they are, most mammals feel reassured when faced by a being who is not obviously displaying antagonism and whose height is suddenly reduced to the level of, or below, their own.

Hunkered on the rock, I continued to speak to the tom. His tail immediately slowed its motion as his ears slowly rose. For a couple of minutes he continued to stare at me; then, relaxed again, he returned to the kill and began to eat anew. But this time he remained standing and glanced in my direction now and then. Satisfied that I had tested him to the full, and now absolutely convinced that he meant me no harm and that he had accepted me as a neutral neighbor, I backed away from the outcrop and turned around, heading home.

As the weeks passed and the winter ripened, we continued to meet, sometimes as a result of my efforts, at other times accidentally, but always peacefully. By mid-December, my knowledge of his habits had grown to the point where I could almost forecast his whereabouts within twenty-four hours of my last sighting, although this did not allow me to see him at will because I simply could not keep up with him over the terrain that he favored. I could follow his tracks readily enough when the snow was fresh, and I could walk the same trails without difficulty, but when he decided to go up a mountain and traveled over steep slopes, or from rock to rock, his agility and surefootedness were such that I was invariably left far behind. After a number of fruitless, dangerous, and frustrating attempts, I learned that when the Ghost decided to climb the mountains in winter, it was useless to try to follow him.

IN THE EARLY HOURS OF DECEMBER 17, UNDER a gibbous moon, I was once again privileged to watch the puma in action. This drama developed in an area at the extreme northern end of the French Creek valley at a place where sedges and willows offered shelter and browse to the herbivores.

I had been following the Ghost for about seven hours, having met him just after 7:00 P.M., when he suddenly entered the trail I was following. He had come out of the heavy forest, in which he had scented game somewhere off the pathway.

His appearance was positively spectral! The trail ahead was empty of all but snow and the shadows of the trees, the night was silent, I was walking slowly, without snowshoes and making hardly a sound while alternating my gaze from his footprints to the way ahead. During one of these shifts of vision I looked down for a second or two, then raised my eyes; and there he was, standing broadside on the trail, gazing at me quietly, ears attentive. He seemed relaxed and his coat appeared almost black against the moonlight. His long tail was undulating, its tip upturned. About fifty feet separated us.

I stopped immediately and spoke to him softly as I squatted and he opened his mouth, his fangs reflecting moonlight. For several seconds we remained thus; then, with a slight wag of his tail, the Ghost turned and continued uptrail. Try as I might, I was unable to hear even a whisper of sound as he walked, his gait elastic, easy, and majestically beautiful.

From then on, I followed him. He moved wraithlike through the night, frequently disappearing from my view as he made use of ground cover or rounded some of the many bends in the forest track. Just as often, he detoured, clearly following the scent of prey, or at least seeking it, returning to the trail when an odor proved to be old, or when nothing alerted his keen sense.

For my part, I kept to the pathway, on some occasions slowing my pace, at others stopping altogether. Now and then I would hear him as he squeezed through underbrush; more often than not, my first awareness of his presence was when he returned to the trail, paused to glance at me, and then continued with his quest. Three times, when some slight noise and a sense of being observed caused me to turn around, I found him behind me, and the hair on the nape of my neck rose stiffly when I realized that I was being stalked— no matter how pacifically—by a formidable, 180-pound predator that could snap my neck with one blow of a massive paw. I was not actually afraid, for my heart did not accelerate its beat and my stomach butterflies remained quiet; yet I was thrilled and excited sufficiently to arouse the neck hairs and to experience a mild feeling of apprehension.

In any event, the Ghost proved that I had no cause to fear him, for each time he entered the trail behind my back, he simply left it again, detouring around me and presently emerging ahead, from there to glance my way before pursuing his own course. In all, I saw him on eight occasions during the time that I followed him that night; most often I lost him altogether as he searched the territory on both sides of the trail or became concealed by trees and shrubs. Nevertheless, that was a memorable experience! The going was rough and exceptionally fatiguing, but the venture was extremely rewarding; it gave me an almost mystical feeling to realize that for the first time in my life I was actually accompanying a large wild carnivore while it hunted its prey. I felt that I was then a part of the Ghost, an extension of this marvelous tawny animal for whom I felt great sympathy now that I had become aware of the enormous challenges that faced him on each occasion that he patiently searched the wilderness for food. I did not see him as a killer; killing was only one momentary aspect of what the Ghost was doing. It was not the end of his quest, but rather the means of it, the first and most basic urge of survival that, if successful, would give him enough food on which to survive one more week before hunger and the vagaries of his life combined to drive him back to the chase. If he failed during two or three such hunts, his chances of staying alive in this pitiless wilderness would be small. His hunger would increase, his energies would begin to flag, his efficiency would suffer; these are the factors that so often lead to death for the predator who must face the intense, biting cold of a northern winter.

Several times the puma disturbed game animals, but they either heard or scented him before he was close enough to charge. I did not see any of this action, but I heard the noise made by the prey as it escaped, the volume of sound suggesting that moose or caribou had been the cat's targets. At these times I could not help wondering whether my presence in the country was responsible for the alertness of the quarry. I had tried to keep far enough back from the puma so as not to intrude on him or on the animals he was searching for, while being able to stay in contact, either by sight, by hearing, or by following his fresh tracks. And I walked as quietly as I could, forcing myself to stalk as though I too were hunting. I don't think now that my presence alerted the prey animals, but I shall never be sure, and that night the doubts came frequently.

By about 2:00 A.M., the tom had led me higher up the flanks of the western mountains, but only about 300 yards above the valley. I was two hundred paces behind him, but had lost sight of him for a moment when I turned a slight bend in the trail and saw him motionless, one front paw raised, his long tail whipping from side to side and his head held high as he concentrated on something that was beyond my powers of detection. Now he cupped his ears and turned his head toward an area of swampland that was below us, and I followed the direction of his gaze with the field glasses. That was when I saw the deer. There were five of them. Four were clustered in one place, the fifth was perhaps 40 yards closer to the creekbank. I estimated that the puma had detected their scent while he was still 150 yards away from the prey.

The night was quiet and made refulgent by the big, misshapen moon; the shadows were darkling forms crisscrossing the snow in open areas, but melding into amorphous, deepcharcoal shapes within the trees and shrubs; a small breeze was coming from the northwest, and the cat, master stalker that he was, had been moving against the air currents ever since I sighted him.

When I saw the quarry, I remained still. The Ghost could not actually observe the mule deer from his own position, for he was lower down the slope, but I was ideally, if accidentally, placed to watch almost the entire action. Each animal, as well as the surrounding lowland, was clearly visible to me. I could not follow the puma's every move because as he advanced he made use of all available cover, going forward at a half-crouch and stopping often to scan the prey with his nostrils and ears. But I knew that by the time he reached the edge of the forest and was close enough to charge the lone deer, I would be able to witness the event.

As the Ghost traveled, his every movement was beautifully coordinated, his great body seeming practically to disappear into the environment. Had I not known where he was, I would not have been able to pick out his indistinct outline, despite the field glasses and the brilliant moonlight.

Before he was halfway to the swamp area, I began to feel concern for the deer. I have often been gripped by that sort of contradiction: I knew that it was important for the cat to make his kill and I hoped that he would, but I felt deep sorrow for the prey and hoped that it would escape. If it were not enough to be so emotionally divided, the clinical part of my mind, the trained biological self, demanded that I should banish emotionalism and concentrate instead on the action, observing with detachment so that not one single element of this rarely observed event would escape my notice.

Several times I found myself on the verge of shouting to scare away the deer, only to stifle the impulse as a sentence formed itself in my mind. *Let it happen. You are only an observer and you must remain neutral!* My hands held the glasses to my eyes, and they saw, my ears listened, even my nose was working, absorbing the odors of the night, the tang of pine, my own scent, and other indefinable essences brought by the breeze. All these mechanical things happened at the same time that my brain continued to argue with itself and my emotions swung back and forth from the deer to the puma.

I had been moving the glasses so that they alternated between the tom and his quarry, but when I lost sight of the cat after it disappeared into the young trees and brush that lined the creekbank, I kept the deer in focus, waiting for the charge that would soon come. I don't suppose that more than a few seconds passed in this way, but the lapse seemed an eternity during which my emotions climbed to their uttermost peak, although my brain continued to function objectively as it interpreted all the messages that my senses were recording. Nevertheless, I felt as if part of me had become disembodied; thus I continued to observe in a calm and detached manner, but at the same time, my blood racing and my heart pounding, I became the hunter *and* the hunted.

When the puma initiated his charge, he broke from hiding so suddenly that the deer didn't have time to react until the puma was halfway to the deer closest to him. Even so, like well-drilled cavalry chargers, the five animals swung away.

As they turned, their forequarters were already lifting preparatory to initiating the first of the series of great leaps that would take them across the swampland, toward the trees on the east bank, even as the puma seemed to be flying over the ground, so prodigious and swift were his gigantic paws.

The targeted deer leaped clear by the merest fraction. I saw the Ghost, his tail streaming like a pennant, reach for the deer's hindquarters with both his massive forepaws, his toes spread and his curved talons seeking a grip. I believe the tips of one set of claws actually scraped the deer's rump, for it faltered in stride, recovered, then leaped, rising more than four feet up and landing twenty feet away, in this manner outdistancing the puma, who could in no way match the speed of such spectacular bounds, even if it were capable of jumping an equivalent distance. The other deer, meanwhile, had disappeared before their companion was halfway to the trees. The puma was outclassed! He leaped mightily four or five times and ran at full speed, but gave up when he saw the target's rump moving farther and farther away.

As the quarry disappeared in the forest, the cat slowed, his mouth agape, his narrow chest heaving as he panted for breath. For about half a minute he stood there, his straining chest and swiftly lashing tail the only signs of discomfiture that he showed. Presently he loped away, following the trail of the deer. But whether he managed that night to down one of them or not, I was unable to learn, despite the fact that I kept after him for another three hours without managing to catch up. As the blue light presaging dawn began to show beyond the eastern peaks, I gave up and started to pick my way back to the creek, descending gradually from the lower mountain slopes until I reentered the valley at a place about half a mile from the lean-to shelter. By now, all the signs proclaimed that a storm was coming. The wind, out of the northwest, was blowing at about twenty miles an hour and heavy clouds were sweeping in. The moon was already obscured, though some of its glow managed to reach the land; the temperature was appreciably colder.

I had planned earlier to spend the day resting in the lean-to shelter I had made in that part of the valley, but as the strength of the wind increased and the temperature continued to drop while the clouds grew larger and darker, I knew that a blizzard was about to descend on me. Instead of seeking shelter in the lean-to, I decided to make a run for the shack, for, without snowshoes, traveling would be difficult by the next day. In any event, there was no way of knowing how long the storm would last. I was on the east side of French Creek when I came to this decision, so when I reached a crossing point that offered dry footing, I climbed up the west bank and picked up my homeward trail.

Dawn had already arrived, but the cloud cover was so heavy that only a minimal amount of light reached the valley and even this was weakened when the snow started. The flakes were not merely falling. They were being *driven* by the continuously blowing wind into which I had to walk in order to reach the valley of the shack. Yet I made reasonable progress for the first half-mile, until the snow became heavier and the winds shrieked themselves into a full northern blizzard. Now, partly shielding my eyes with an open hand, I struggled along at a snail's pace, trying to stay on the trail, but often turning away from it when blinded by the stinging snow. Although the light was stronger after about an hour, I had no means of knowing just how far I had progressed. I could not become hopelessly lost because of the guiding creek, but I could not distinguish any landmarks. And I was getting cold.

I believe I spent another half-hour stumbling along in this way when, tired of being buffeted by snow and wind, feeling now *intensely* cold and almost blind to my surroundings, I decided to seek shelter. To this end I climbed a little higher to try to get into the lee of the trees while looking for a rock overhang that might be snow-filled and deep enough to tunnel into. Twice I thought I had found such a shelter, but investigation proved each one to be too shallow for my needs. Then, as I was starting to feel somewhat desperate, I spotted a craggy overhang that bulged with snow.

I began digging a tunnel into the drift, and the deeper I went, the more encouraged I became, for such a shelter can be made windproof by covering the entrance hole with snow, and the inside temperature can be maintained at a comfortable level with but the heat of one candle, six of which I carried in my pack. Here I would have light and heat and enough food to wait out the storm.

I redoubled my efforts and was pleasantly surprised when my right hand broke right through the snow and penetrated a space. Unknowingly, I had found some sort of cave, an *ideal* shelter. In moments I had made an opening wide enough for my head and shoulders. Despite the poor light, I could now see what I was doing, for there was no snow to get into my eyes or to interpose itself as a curtain before them. As I tried to crawl through, my backpack became jammed against the roof of the tunnel I had made, so I retreated until I could reach up and scrape away more snow to make room, then I crawled forward again, wriggled, and pushed the upper part of my body into the somewhat rank-smelling cave, the interior of which was just a few shades away from pitch black. To get more light, I made the entrance wider. That was when I saw the bear.

The bulk of an enormous grizzly appeared as an indistinct mass that was somewhat lighter than the walls and sloping roof of the cave. The animal was lying at the end of a chamber that was about eight feet wide by about ten feet deep. The roof was no more than five feet high at the entrance, then shelved to less than three feet. The grotto was part of a granite overhang, a natural concavity of the kind often encountered in mountain country; indeed, it was the sort of shelter that I had been hoping to find—but without an occupant!

The bear lay like a shaggy, bunched rug, his great head resting on both massive forepaws. He was facing me. Except for the slow heaving of his lungs, he wasn't moving, but his eyes were open and they stared balefully into mine. Incongruously, I remember thinking that I had never before noticed that grizzlies have such small eyes! After that, for a few eternal seconds, I just lay there, thoughtless and immobile, my head raised, my mittened hands pressed flat against the cold floor of the den. But when the bear lifted his head and uttered an explosive hoarse grunt, my body became electrified.

Powered by an instant flood of adrenaline, I popped out of the cave like a champagne cork from a bottle. Outside, I jumped to my feet, grabbed the backpack from where I had put it before tunneling into the snow, and ran downslope as fast as my legs could manage. Slipping and sliding, jumping over icy rocks and dodging trees, I raced down to the creek without knowing whether I was being chased or not. Then I glanced back. I *was* being pursued! The enraged grizzly was about seventy-five feet behind me. Reflexively, I dropped the cumbersome packsack, which I hadn't had time to shoulder as I decamped from the den, and now, as fearful as I have ever been at any time during my life, I redoubled my efforts while I searched frantically for a tree that would be easy to climb. After some moments I couldn't resist another backward glance, and my relief was great when I saw that the grizzly had stopped chasing me. He was still angry, but he was venting his justifiable displeasure on the snow and on nearby bushes, swatting left and right with his front paws and causing green and white debris to rise briefly until it was carried away by the storm. These outbursts of temper were accompanied by small jumps that raised his forequarters at those moments when he swatted with one or the other paw while he swung his head from side to side, glaring at me.

I slowed, but kept glancing over my shoulder as I made for a straight lodgepole pine that was small enough to climb quickly. Beside the tree, I faced toward the grizzly, a big male who was now about 125 feet away and seen as a snow-shrouded figure standing with his oversized rump to the wind, head held high and nostrils flaring. He had evidently lost my scent, which was not surprising, for the storm was blowing fiercely and toward me. By now I was in control of myself. I knew the bear would not pursue me any further, and that he would soon go back to his rock den. But I remained tense until he turned around and started to walk upslope, soon to disappear inside the forest.

I had been lucky, first because the grizzly had been befuddled by sleep, second because I had reacted so quickly, and third because the animal was clearly discomfited by the storm and anxious to return to his shelter. I had also been careless. The storm had developed into a full blizzard and I had deemed it prudent to seek shelter, but my anxiety to get into cover caused me to enter the grotto without first making sure it was unoccupied. As a rule, a bear's winter den is distinguished by a yellow rime encrusting a small hole located somewhere at the top of the entrance, a discoloration resulting from the animal's exhaled breath, which is always warmer than the outside temperature and melts an escape route for itself. The storm had packed new snow over the breathing vent, something I should have expected and been looking for. Had I taken the precaution of brushing off an inch or two of snow, I would have encountered the crystallized yellow area.

Reassured by the grizzly's departure, I nevertheless waited beside the lodgepole pine for fifteen minutes, wanting to be certain that he had, indeed, returned to his bedchamber.

And as I stood there buffeted by the wind, I gave thanks to the gods for allowing me to emerge unscathed from an encounter that could so easily have ended my life. I shuddered at the thought, not so much because of the manner of such a death, but because my ultimate fate would have remained a mystery. Once I have quit this planet, I care not one whit about the leftovers and I wouldn't mind a bit if they were put to good use by the animals of a northern forest, but I would like my passing to be noticed by at least one other human being.

Presently, ridding myself of such macabre thoughts, I went to collect the discarded packsack, determined now to ignore the blizzard and to stop only after I reached the shelter of my own home.

What a trying journey that was! The cold had intensified, the wind was blowing about fifty miles an hour, shrieking down the valley and carrying before it the lashing, blinding snow. Fortunately the storm was at my back, or I would have *had* to find some sort of shelter; as it was, I climbed up the bank, seeking cover among the trees, but then I found that visibility was reduced to a matter of a few feet in any direction.

Without realizing it at first, I wandered off the trail and blundered farther up the slope than I intended, becoming aware that I was at least temporarily lost when a blown-down pine barred further progress. But the furious wind was itself a guide. It continued to funnel southward, combining with the slope of the land to lead me back into the valley, until I again found myself on the trail; from there it was a short walk to the creekbank. I kept to this guideline, fearing more serious difficulties if I again missed my way, but enduring considerable hardship as I made slow progress.

The blizzard seemed bent on reducing visibility to zero. Not only was it driving the snow before it, but it also picked up the loose flakes from the ground and from the trees, and these swirled like thick fog, producing a dancing white curtain that opened suddenly at times to reveal the way ahead, then closed as quickly. I was not cold, being well dressed for the weather and additionally warmed by exertion, but I was tired and hungry. It was impossible to measure my rate of progress because landmarks were totally concealed, and if it had not been for the partly frozen creek beside me, I would have been lost an hour after seeing the last of the grizzly.

Despite the fact that I had been moving so slowly and seeking to stay in touch with my surroundings, I almost missed the trail to the valley of the shack, noticing it at the last moment when some instinct caused me to hesitate, then stop, just as I had passed the opening. Never was I happier to see the ugly little shack. And never was I more grateful for the heat of the stove, and for the hot coffee that I drank within twenty minutes of entering my shelter.

The storm grumbled itself into silence sometime during the night, and by morning the sky was clear and the sun was again inching toward the peak. I got up late, after seven o'clock, added fresh wood to the stove, put on the coffeepot, and dressed. While the water was heating, I went outside to get a fresh supply of logs and to look at the new day. The eastern peaks were silhouetted by a deep orange glow and my valley, lying westward, was bathed in rosy hues. The snow was deep; it reached above my knees near the cabin and was undoubtedly deeper than in open areas. Wisa, Ked, and Jak flew to me as I was about to reenter the shack with an armful of wood, so I left the door open and they followed me inside. After stacking the wood, I turned to find some food for the three birds, who were walking about on my clumsy tabletop, cooing and whistling, asking me to hurry up. When they had stuffed their beaks and gullets with trail rations and left to hide the spoils in the forest, I put a supply of food outside the door for them, filled a galvanized pail with snow, and put it on the stove to melt and heat for my morning wash and my weekly stand-up bath. Meanwhile, I prepared breakfast and made coffee.

When I was ready to leave for one more day of puma tracking, this time planning to use snowshoes, it was almost 10:00 A.M. The sun was just above the eastern peaks and the wilderness was again a place of magic and infinite beauty. Outside, mushing over the wind-packed snow, I walked toward the Goldstream River valley, which I proposed to explore from the place where it was intersected by the French Creek valley, on this occasion planning to go west at least as far as the pond where I had seen the owl during my journey upriver. This is the widest and flattest part of the entire area, encompassing country that is in some places well treed, in others more open and marshy. At the sandbar where French Creek develops three tributary arms in addition to the main stream, I found that the water was still running slowly, but that the borders of all streams were frozen. Crossing the open area here and stepping over the creek at a place where it had become narrow now that the runoff water had almost ceased to flow, I walked toward the Goldstream in hopes of finding a place where I could cross it without getting wet. In this I was disappointed. Nowhere had the river narrowed sufficiently to allow me to leap across, but at one place, almost directly in line with the sandbar, there was a small island in the center of the waterway. An eight-inch-thick dry lodgepole standing nearby served as a portable bridge, after I chopped it down with the axe, trimmed its top, and manhandled the last twenty feet of trunk to an upright position right on the riverbank. When I let the pine fall, pushing it toward the little island, it just spanned the open water. With a ten-foot balance pole, I crossed over, dragged the pine onto the island, wrestled it to an upright position once again, and repeated the performance. At the end of thirty minutes' work,

I stood dry-shod on the south bank of the Goldstream and walked across an area that was somewhat marshy, but frozen and covered by three feet of new snow. Gaining slightly higher and firmer ground, I paused to examine my surroundings, spending a few moments admiring the sparkle of Downie Peak, an ice-crowned spire and the dominant feature of the country thereabouts, that stands 9,607 feet tall. Located some nine miles east of where I stood, Downie, I knew, was practically surrounded by icefields, the largest of which was one and a half miles long by one mile across at the place of its greatest width.

Soon afterward I walked along the edge of a dense forest of evergreens that marched resolutely up the mountain to within a couple of thousand feet of the inhospitable, barren peaks. I was heading almost due west, the Goldstream located north of me, but often hidden from view by aspens, water birches, and some great cedars. Here and there, on higher ground that rose out of the valley, clumps of tall, arrow-straight hemlocks grew. Tracks were profuse in this area and included those of moose, caribou, and wolves, indicating by their fresh impressions that the animals had passed this way perhaps two hours before me.

Because of their longer legs, the deer and moose had found firm footing without sinking into the snow, but the wolves had gone in belly-deep, the indentations made by their bodies clearly evident. Under such conditions, prey animals can move more freely than hunters, which are forced to leap continuously in order to travel through the deep and heavy snow. This is exhausting work and the big predators often go hungry until a thaw comes to put a hard crust on the surface; this turns the tables and gives the advantage to the predators, for while *they* can now run on top of the snow, hoofed prey animals break through it, often cutting their legs on the sharp edges of the crust.

There was no sign of the puma, but I had not expected to find any in this area, which I believed to be the extreme northeastern boundary of the female cat's range.

By lunchtime, although I had enjoyed the leisurely, pleasant journey, I had made no new discoveries. While I was resting, a pair of pileated woodpeckers flew over my head as they traveled eastward above the Goldstream River, their flame-red heads flashing in the sunlight, their bodies rising and dipping in unison; in moments they angled northward and disappeared into the forest. I was sitting on a downed log, drinking vacuum-flask coffee soon after the woodpeckers entered the shelter of the trees. A few minutes later a barrage of staccato blows rattled across the wilderness to proclaim that both birds were hammering at dead trees to get at the dormant insects that were sheltering deep inside the wood.

I took out a bannock-bread sandwich, made from half of the contents of one of the precious cans of corned beef. It was frozen hard and had to be chewed slowly if I was to taste it. I was forced to share some of the treat when a gray jay flew into a tree immediately near my seat. The bits of bread I threw were quickly retrieved, and by the time I had finished the first half of my lunch, the bird was at my feet, waiting to take bits of meat and bannock bread from my fingers. In the meantime I was kept constantly entertained by the chickadees and nuthatches, about a dozen of the former and three of the latter, the little so-called red-breasted kind that are really a deep chestnut underneath. Above us all, circling tirelessly in the thin air and contrasting with the blue of the sky, three ravens eyed the forest floor, looking for their own lunch. Around me, the delicate trails of white-footed mice were plentiful, the marks of their pigmy feet pressed lightly in the snow and followed by thin lines made by their dragging tails. They were evident wherever a mouse had popped up out of one of the many tunnels that these pleasant little animals build under the surface.

After lunch, nearing the full skirts of a tall cedar, I found a number of marks that combined to tell me about a wilderness drama that had occurred during darkness, but after the blizzard had ceased to pelt snow at the land.

A great horned owl, seeking to satisfy its hunger after being confined to its roost by the storm, had gone hunting when the wind was stilled. In the same neighborhood, but underneath the snow, a ruffed grouse had sat out the storm, warm and secure within its burrow until something had startled it and caused it to burst out to try to seek safety in a tree. The owl had struck. The great hooked talons stabbed into the bird before it had traveled ten yards from its bedding place. The grouse was plucked and consumed on the spot.

It is easier to tell this story than to describe how it was pieced together, yet this should be explained, because the reading of the marks left by the animals that dwell in the wilderness plays a major role in the study of life. Each species that passes through the forest or over the fields and deserts leaves its own particular signs, and these, either individually or collectively, can be looked upon as the "writing" of the wild, a record that is often more interesting and informative than the sighting of the animals themselves once it is understood, because it frequently tells the full story of an event.

When my attention was drawn to the area of disturbed snow near the cedar tree, I immediately became interested and walked to the scene, where I stopped and looked closely at the marks. Clearly noticeable were two holes in the snow: The first of these was more of a slanting tunnel that began as a shallow furrow and ended at the place where something had burrowed down into the white canopy; the second cavity was some five inches deep and seven or eight inches long, situated eighteen inches in front of the first. Around this second hole, particles of crystallized snow glinted in the sunshine, icy crumbs varying in size from rice grains to dried peas, which had been dislodged by an outwardly directed force. Inside, the walls and bottom of the furrow were smooth and frozen; its former occupant had left in the hole a small pile of yellow-gray droppings, each about an inch long. This completed the first "chapter" of the story.

From experience I knew at once that this had been the bedding place of a ruffed grouse. On particularly cold nights, or during storms such as we had experienced, these woodsy-colored birds dive into the snow, tunnel a short way under it, and settle themselves down for the night. If undisturbed, they emerge at dawn and go about their business, but if startled, they burst out of these "igloos" and usually whir away safely to seek shelter in trees. Occasionally, when there has been a big thaw during the day and this is followed by a really cold snap before dawn, a grouse may become entombed in its bedroom when the wet, soft snow hardens into a thick crust. If this happens, the bird either freezes to death and is later eaten by a scavenger, or is eaten when a predator's sensitive nose and ears lead it to the struggling captive. (Once I inadvertently freed such a prisoner when I stepped on it with the front of one of my snowshoes; the bird's sudden, explosive exit gave me a first-class start.)

The difference between a leisurely departing bird and one that has been startled out of concealment is readily seen around the hole; if the grouse has emerged to seek breakfast, the surface around the exit is crumbled and many of the "crumbs" fall into the cavity, but when the departure has been hurried, the debris flies outward, becoming scattered outside the hole. So a quick glance told me two things: A grouse had bedded down, and it had been suddenly disturbed.

Confirming the identity of the victim and the success of the attack was a pile of feathers about twelve feet from the exit hole, evidence that was additionally supported by a few drops of blood and pieces of bloody skin from the victim's body. The snow in several places showed the impressions of flight feathers: sculpted, fanlike designs with softened edges that could only have been made by a large owl. This narrowed the identity of the hunter to three species: the great horned owl or the great gray owl, both natives of the region; or, less likely, the snowy owl, a denizen of more northern latitudes that sometimes winters in relatively southerly areas. These three are the largest owls; one of them had left the widespread wing tracks.

One significant fact pointed directly to the great horned owl. Since the grouse was literally buried under the snow, none of the birds would have seen it. By the same token, the grouse could not have seen the predator; neither could it have heard it in flight, because owls are silent fliers by reason of the fact that the margins of their feathers are naturally "teaseled," softened in such a way that they muffle the sound of the air they displace. Yet *something* had panicked the grouse, and the absence of any animal tracks in the area precluded the possibility of its having been startled by a land animal.

What, then, had caused the grouse to break cover? Knowledge of *Bubo virginianus* supplied the answer. I was quite familiar with the great horned owl's sudden frenzied screech, emitted full-voice for the express purpose of startling possible prey out of concealment. The demoniac scream is used by this efficient predator for two practical purposes that I know of. In the first case, the owl screams when it has noted indistinct movement in the night and is not sure of the identity of the animal that has made it; it might be a hare skulking in the shelter of some bush or tree, or it might be a fox or a bobcat. So the owl launches its banshee wail suddenly and piercingly; it would take nerves of steel to endure it without flinching. Prey animals panic and run; predators are startled sufficiently to reveal themselves.

In the second case, the results are the same, but the reason for the scream is different; in this instance the owl hasn't seen *anything* for some time. It is hungry and impatient, so it screams in the hope that the bloodcurdling yell will flush out some quietly crouched, prospective victim.

That was what must have happened to the grouse. The owl, perhaps perched in the nearby cedar, had surveyed the moonlit valley, sighted no food, and then screamed. The snow-covered bird below had panicked and scrambled out to its death.

This was a "book" that I had ready many times in other places. It took but a few glances to decipher the story before I passed the scene of rapine and veered away from the creek to enter an area of heavier timber. Here the snow wasn't so deep, for the full skirts of the evergreens kept out the drifts. Half an hour of quiet walking took me into a region that had once been swept by fire and was now naturally reforested by second-growth balsam firs, cedars, and aspens. Here I found the marks made by the passage of mule deer. The signs could hardly be called tracks because the animals had sunk into the snow up to their bellies and had been forced to travel in leaps and bounds. The condition of these marks suggested that they had been made several hours ago, and when I stopped to examine one of them, I noticed a scattering of dark droppings that had already "burned" themselves downward by absorbing the heat of the early sun and melting the snow around them.

I stood quietly for several minutes as I surveyed the tangled landscape and listened to the tempo of the wilderness—the bird sounds and the small noises made by a gentle breeze stirring the tops of the trees. Looking and listening, I remained unmoving for a while before noting the position of the sun and deciding it was time to go on. But at that moment

I heard the sound of a heavy animal moving through the snow to the north of where I stood. It was coming my way. A deer, I thought, as I took two very cautious steps so as to place myself silently against the lower branches of a cedar. I waited expectantly. The animal was moving slowly and stopping often. The sound of its passage came from about seventy-five to one hundred feet away.

Soon afterward, the unseen animal angled toward my right, evidently moving toward a small open space about thirty yards from me and in full view from my position. A few moments later I detected movement on the far edge of the clearing, but still within the last screen of trees and bushes. Continuing to believe that it was a deer, I did not try to conceal myself. The animal was now remaining still, no doubt listening and smelling the air before venturing into the open. Then, with unexpected suddenness, a puma emerged, stepping out of the forest and striding purposefully into the clearing, belly-deep in snow.

It was not the Ghost. This cougar was at least 25 percent smaller and had a finer, slimmer head and darker shadings around the face. For a matter of seconds it stood still, smelling the environment, listening, scanning carefully with its eyes. Then it turned to face me, having to stand on its hind legs to complete the movement and thereby allowing me to see that it was a female, almost certainly the cat whose tracks I had found in the vicinity of Goldstream Creek. These things I noted in a few seconds while I waited for the animal to dash across the opening and disappear in the forest. But she didn't move. Instead she focused her eyes on the cedar next to me and aimed her eyes in my direction, her nostrils quivering as she sniffed. Whether or not she knew I was human, that puma knew that some kind of living thing was located beside the cedar. The fact that she did not run away made me wonder about her intentions.

Less than a hundred feet separated us. This meant that despite the absence of ground-level wind, she almost certainly had my scent. If she was the nervous kind, this alien smell should have been enough to send her running for cover, so I was forced to conclude that she was bold and therefore unpredictable.

For years I have made a practice of sewing a large, wide pocket in the interior of my parkas so that I can carry things that require protection from the elements, or are delicate and must be guarded against damage. In grizzly country, I invariably carried three flares, the kind used by motorists to signal an emergency. Contained in a leatherette pouch, these signals, when ignited by a pull-tab, are calculated to discourage a bear if used at first signs of aggression. Until I blundered into the grizzly's den, I had never needed to make use of the flares—and during that encounter I just didn't have time to use them; nor would it have been sensible to employ one in such close quarters! Now I was glad I had kept them in the parka pocket, for it seemed that the female puma was disposed to attack.

Because I was hot from walking, my coat was unbuttoned, so there was no need to fumble with the fasteners. I was slowly reaching upward with my right hand when the puma cut loose with a bloodcurdling shriek, a great banshee scream that must be heard to be really appreciated. I had heard lions scream on several previous occasions, but always from afar; nothing I had ever listened to before, or have heard since, can compare with that frightful cry. The urge to run was so strong in me that I almost gave way to it, especially when the unearthly yell rose to a crescendo, stopped suddenly, and was at once repeated.

As I struggled with rank fear, I continued to stare fixedly at the cat—not because I *wanted* to, but because I couldn't help myself. She advanced slowly as she yelled, stopping when the first scream came to an end and the second was unleashed. Her ears were flattened against her skull, her tail lashed viciously, and her eyes blazed with what I interpreted as fury; her mouth, meanwhile, was fully open to reveal an array of teeth framed top and bottom by the great fangs.

Three times the lioness shrieked while continuing to advance slowly, somewhat hindered by the deep snow. I have no idea how long I stood immobile, though it couldn't have been for longer than a few seconds, but the interval served its purpose. By giving me those few moments of grace while she repeated her dreadful cries, the puma allowed me to recover my wits. I grabbed for the flare envelope, pulled out one of the signals, ripped off the tab, and as it burst into intense, ruby flame, I threw it at the lioness just as she was charging.

The first flare was still arcing through space when I had the second one in my right hand, the thumb and forefinger of the left clamped on the pull tab. The bounding cat and the flying flare nearly collided about forty feet from where I stood, the burning signal hitting the snow a few inches in front of the charging animal. I don't know whether it was the intense light or the hissing sound of the burning flare that caused the cat to turn away. But I was greatly relieved when she stopped suddenly, her snarls loud, and swung herself to the right. She started to back up, but she didn't run away!

After moving backward for about ten feet, she stopped, still facing me. She continued to snarl malevolently, her ears pinned back and her tail lashing wildly as she crouched, her exertions causing her to sink belly-deep into the snow. Her long tail was cocked upward, sweeping across the top of the snow and creating a slightly depressed, fan-shaped drag mark. But it was the eyes of the animal that impressed me most. Although they were wide open, the wrinkles created by the gaping, snarling mouth had traveled upward so that the loose skin beneath each eye socket caused the bottom lids to cover the lower half of her amber orbs, giving them a sinister appearance even though they seemed to be looking at me more in appraisal than in real anger.

As the flare hissed and sputtered and slowly sank into the snow, creating a rather weird, incarnadine glow around the place where it had landed, and as the puma continued to stare at me and snarl as vehemently as before, I regained full control of myself. Objectivity replaced emotion and I began to analyze the situation.

The cat was unquestionably dangerous—lethal, in fact— but her charge had lacked the speed and purpose of those made by the tom, whose aggression I had twice observed in detail. Had she come at me as suddenly and as swiftly as the Ghost had attacked the deer and the caribou, I would not have had time to get the flare out of its case, pull the tab, and throw it. These things brought to mind the occasions in the past when I had been threatened by black bears, which employ what I call a bluff-charge when they are approached suddenly and feel that they do not have time to flee. Because I had studied bears for many years and had seen them initiate their half-charges at other bears and especially at my sled dogs, I was prepared the first time that one of them did the same to me. Holding my ground and clapping my hands while talking to them in a loud voice, I had on seven occasions outbluffed lone bears, all of which had slowly retreated until they felt that it was safe to turn their backs on me and run.

A puma's charge differs from a bear's only in the manner of its execution: The cat is faster, its leaps cover more ground, and its method of attack, once it has reached a quarry, displays a deadly, coordinated finesse that is quite lacking in the brute-force approach employed by a bear. Both animals are equally dangerous, but the puma kills with the dexterity of a rapier, whereas the bear downs its prey with the laborious strokes of a broadsword.

The cat's noisy and frightening advance had lacked the determination and purpose of a real, committed charge. Only a few seconds elapsed between my throwing the flare and the animal's retreat, but the respite allowed me to control my emotions sufficiently to realize the difference. The cat, meanwhile, continued to crouch and voice her noisy protests without seeking to initiate a second attack, giving me a little more time in which to decide on a course of action.

When one is in the open and faced by a potentially dangerous animal that can run at speeds much greater than one's own, it is useless to seek to escape to safety. Panic, being a communicable emotion, serves only to encourage an attack; predators instinctively chase animals that attempt to run away, just as they can immediately detect fear and become emboldened when it is displayed. Reason and strength of will are the only effective weapons possessed by an unarmed human who is faced by the kind of threat that crouched before me that morning.

The cat began to move sideways, seeking to approach me from the left, where I was unprotected by the trunk of the cedar. As she did so, I took a step in her direction, addressing her in a loud, controlled voice. She stopped immediately and crouched down again. At this point, less than thirty feet separated us. I held the second flare; the first was still burning under the snow, but its light was barely visible, and since I noticed that the cat's movements had taken her away from the hissing signal, I took another step and turned slightly so that I was aiming at the flare. The puma retreated, moving backward, but she stopped almost at once. Her snarls had become lower, but her fangs remained exposed and her ears flattened. I couldn't see her tail because this had been forced down under the snow, but I presume that she was still trying to lash it.

I was now convinced that the cat was reluctant to force a confrontation, but I was greatly puzzled by her behavior. I had not come upon her suddenly and there had been ample opportunity for her to escape into the forest after she first detected my presence. Why, then, had she decided to approach me and to initiate a halfhearted charge? At the time I was not disposed to try to find an answer. Instead, displaying a boldness that I was far from feeling, I tramped toward her in a manner that I fervently hoped was intimidating. Each time I took a step, I lifted my foot high so as to display the long, oval, flapping snowshoe, thus introducing her to a completely new experience. I made as much noise as I could manage, talked incessantly in a gruff voice, and flapped my arms.

I had only taken three steps when the cat broke. Snarling one last time, she backed away, turned, and bounded across the small clearing to disappear into the trees. I was just heaving a sigh of relief when she screamed again, the long, unearthly shriek seeming to fill the entire wilderness. Then there was silence, a complete cessation of all sound, as though every animal and bird in the vicinity had been struck mute by the great primordial cry. The only noises to reach me during the next five minutes were the soft, whispering murmur of the breeze as it traveled through the tops of the evergreens and the faint hissing of the flare, which had by now sunk right down to the forest floor.

While I waited to make sure that the unpredictable cat had really abandoned the area, I gave more thought to her behavior. Pumas are usually silent animals. Except for their snarls when charging or fighting, they seldom vocalize beyond purring when they are content. At mating time, however, the females are known to scream loudly, sometimes for prolonged periods, while the males, attracted by the female's wail, are given to uttering deep, persistent whistles as they approach the cat. Had I unwittingly placed myself in the path of a female lion who was even then entering a period of estrus and was out trying to attract the attention of a tom? I concluded that this was the only logical explanation for her behavior. Those people who have been prevented from sleeping by the amorous cacophony that rises when two domestic felines are engaged in mating are aware that cats behave quite irrationally (from a human point of view) when they are dominated by that one primitive urge. Even the most home-loving, mild-mannered cat undergoes a complete personality change when she or he is enraptured in this way. The pre-mating ruckus that follows has caused more than a few people to rage when they are awakened by it in the small hours of the night. Imagine, then, the kind of racket that can emerge from a female puma when she is aroused by the mating urge!

Pumas have no regular breeding season. Unlike most other large predators, the females are polyestrous; that is, they come into heat more than once a year when they are not carrying young, a characteristic shared by most felines. Sometimes they even go into estrus soon after giving birth to a litter, but if they mate at this time, it appears that the development of the fertilized eggs is prevented by the need to produce the milk for their offspring. By the same token, it is not known how many times a female cougar will come into heat during any one year, although it has been definitely established that when a period of estrus is not consummated, she will develop another later on. For this reason, kittens may be born during any month of the year, although late winter and early summer appear to be peak times.

Between ninety and ninety-six days after fertilization, from one to six kittens are born, the more usual number being between two and four. These stay with the mother for about a year, which means that a cat does not produce young every year. Gestation times vary, estrus periods are unpredictable, contact with a male is not always certain during each period of heat, and availability and abundance of prey often determine the amount of time during which the young are allowed to remain with the mother.

After waiting for ten minutes, while the birds and squirrels in my vicinity resumed their normal pursuits, I left the shelter of the cedar and continued westward for about a mile before stopping to have lunch. Near the partly frozen river I selected an open space that was in full sunshine and gave me a good view in all directions. As I ate, I continued to muse about the puma's behavior and reviewed in detail my knowledge of the species. At the end of an hour, my fright was replaced by anger at myself for displaying weakness. And I was determined to learn more about the she-cat. If I was right and she was, indeed, in heat, this would surely mean that the Ghost was soon going to become involved. To the best of my knowledge, he was the only male in the region.

I returned to base that afternoon to equip myself for a longer stay in the female's territory.

AIDED BY MOONLIGHT AND UNDER CLEAR skies, but in temperatures that were constantly well below zero, I spent a number of nights exploring the female puma's territory. I was rewarded by hearing her scream on four different occasions, each outburst voiced between two and six times. To listen to such cries in the dead of night in wilderness country, when there is no other human being within thirty or forty miles, is an experience that causes the emotions to oscillate like the prongs of a tuning fork. Every time the animal called, I found myself initially gripped by intense apprehension, but this soon turned to fascination and as quickly led to exhilaration. The puma's wails were at the same moment supernaturally terrifying, ineffably *wild*, and utterly exciting. The only comparable sounds that I know are the howls of timber wolves or the tremulous and plaintive wails of loons calling from some night-shrouded lake.

At 1:00 A.M., during my last night in her range, the female cat called while we were near the entrance of the Goldstream Creek valley, separated from each other by about a quarter of a mile. On this occasion she repeated her cries four times; the third and fourth calls, which I timed, lasted 5.69 and 7.53 seconds respectively.

The moon cast good light that was augmented by refraction from the snow. These conditions allowed me to move relatively quickly as I half-trotted down the creek valley in the hope of finding the cat's recent signs. For two hours I followed her, first picking up her tracks about half a mile from the pond; but she managed to stay ahead of me throughout that time and eventually defeated me altogether when she turned away from the valley and began to climb up the westernmost slopes of Downie Mountain, an area that was much too steep for me to follow.

What was left of the remainder of that night I spent in the lean-to shelter I had built in this part of her range, rising at dawn to resume the search after breakfast. All day I sought her tracks without success, but by late afternoon, having returned to the river valley and followed it eastward, walking on the south side, I again crossed her tracks. She had evidently climbed some way up the mountain, traveled northward, and descended to the lowlands at a place where there is an aging log bridge that allows dry passage over Brewster Creek, a waterway that tumbles down from its source among the Downie icefields. From here, during the early hours of morning, she had followed the river valley in an easterly direction, keeping between the water and the sloping land.

By early evening, in the area of the swamps that lie about a mile west of the mouth of French Creek, I found that the puma was literally walking in my outward-bound trail, no doubt because my snowshoes had compressed and hardened the snow, which now offered firm footing, and had continued to walk over my tracks right up to the three trunks I had placed as bridges between the small riverine island and the bank, so I could cross without having to get wet. (Recently I had cut and positioned a second tree, thus saving myself the trouble of moving the first one each time I needed to negotiate the water.)

The puma had used the makeshift bridges to cross the river. She appeared to be heading toward French Creek, a route that did not greatly surprise me because by now I was convinced that she was, indeed, gripped by the breeding urges. With the complete honesty of the wild being, she was looking for a suitor, undoubtedly well aware that the Ghost was to be found somewhere within the territory that lay north of the Goldstream River. In all probability, the two pumas had met and mated in the past. It appeared as if they were about to repeat the experience.

Picking up the balancing pole that I had left on the south bank near the tree-trunk bridge, I removed the snowshoes and strapped them on the outside of the packsack. Crossing the river took only minutes; on the far side I again strapped on the snowshoes and picked up the cat's trail. As I had surmised, it led me directly to the sandbar, but whereas I had expected the female to follow the French Creek valley, she had instead followed my own trail, the one that led to the valley of the shack.

Although my comings and going along this route had compacted the snow, the dimpled patterns left by the webbing of the snowshoes allowed me to follow the cat's spoor with comparative ease because the weight of the animal had flattened a number of web-lines wherever she set down a paw, though the complete outline of each track was never clearly evident.

Arriving at the trail mouth, I saw that the puma had circled my small dwelling and then examined the wreckage of the old mine building before continuing northward, staying within the trees, but walking near the bottomlands at an elevation that was slightly more than two hundred feet greater than that of the waterway.

Dusk was moving in and the skies were clouding over, the wind blowing from the west at about twenty miles an hour. The moon would be obscured that night and snow was almost certain to fall within the next few hours, so I decided to spend the night at base and set out early the next morning. In any event, now that the cat was seriously searching for the tom, I thought it would be better to allow the two to make contact undisturbed. The excitable female might be less aggressive if she met me again while in the Ghost's company.

Details relating to the courtship of pumas are rather meager, but the fact that this is usually initiated by the female has been pretty well established. Less certain, but supported by a fair amount of evidence, is the interval during which male and female live together once they have made contact. This period varies from ten days to three weeks, the time limit evidently depending upon hormonal changes in the female after she has become impregnated. Once the fertilized eggs begin to divide, the cat's estrus ends and her hormonal balance is altered by the demands of pregnancy. When the female reaches this stage, she is no longer interested in the male. She now rejects his advances and begins to travel back to her own range. The amorous tom, whistling plaintively, follows for a time, but after having his ears boxed and his nose scratched by his former mate, he eventually gives up and returns to his own affairs. Up to that time, I had only heard the row that develops when a cat is engaged in discouraging a persistent tom, but the snarls and yowls produced by such a fracas carry over considerable distances. A listener might well be led to believe that the pumas are killing each other, for the noise is ferocious, but if the male sustains injuries, these are minor. At worst, a tom abandons his pursuit after receiving a few slashes on the nose and muzzle and perhaps having one or both ears slit, but since such auricular cuts are made in tissue composed largely of cartilage, there is little bleeding and the wounds heal quickly, although the tom is forever after destined to display tattered ears because the cut ends of cartilage will not become joined by the natural healing process.

Thinking about these things that evening while I was preparing supper, I fervently hoped that I would have an opportunity to observe at least part of the courting rituals of the lions.

The next day was December 24, the significance of the date failing to register until after I had finished breakfast and had already stroked off the numbers on the calendar. Outside, the snow was falling steadily, but there was no wind and the temperature had warmed up, now registering fifteen degrees above zero Fahrenheit. Not a bad day at all in which to continue my task. But when I at last realized that it was Christmas Eve, and in view of the fact that I had not enjoyed one full twenty-four-hour period of rest and relaxation since September, I elected to have a small celebration of my own. To lessen the feelings of guilt that assailed me immediately after making the decision, I persuaded myself that the time of idleness in which I was going to indulge would give the pumas a chance to become better acquainted. It may be that I salved my conscience by resorting to sophistry, but as matters turned out, the stolen holiday was to yield some excellent, if unexpected, results.

In the absence of other humans toward whom I could show goodwill, I prepared a special treat for Wisa, Ked, and Jak: a mixture of dried fruit, raisins, nuts, and roasted soybeans, which I carefully arranged in the lid of one of my cooking pots. The final result was a circle of fruit within which was a circle of raisins, then a circle of beans, and finally a solid, two-inch bull's-eye of nuts. A most festive plate!

I put the food on the table, stoked the heater full, then opened the door, intending to whistle for the birds. They had anticipated me. The three were perched on the branches of a young lodgepole that stood just outside the cabin door, and the moment I appeared, they launched themselves at me. Wisa landed on my right shoulder, Ked alighted on my head, and Jak came to rest on my forearm. I wished them all a merry Christmas and took them inside the cabin. With cooings and soft whistles, the birds flapped onto the table where, despite the season, they immediately engaged in a squabble initiated by Ked, who was the boss bird. In the excitement of the moment, each jay managed to tramp into the neat arrangement, some of which was scattered on the tabletop. But I didn't mind. I was glad to have some company with whom to share the day.

For an hour the three jays made incessant journeys to the food, sometimes actually swallowing a nut or two, or some fruit, before filling their mouths and gullets and streaking away to hoard the supplies in the forest. When the lid was practically empty and the last bird had crammed his beak and flown away, I closed the door, put two more logs on the fire, dressed myself for the outside, and went to join my companions. As I bent to secure the snowshoes to my boots, the trio returned and landed on my back. Wisa, as though to wish me well, nibbled at the lobe of my right ear.

I walked and the birds either perched on me or flitted back and forth between food-storing journeys, for I carried a pocketful of trail rations to share with them, keeping within the trees in the immediate neighborhood of the shack. The snow was drifting down gently; big flakes, like the downy feathers of a white swan, each distinct and lazy. The surface was entirely covered by the new fall, presenting a pristine, contoured panorama; the trees were freshly dressed, each resplendent in its individually tailored coat, and if the skies were cloud-laden, each light gray mass was singularly formed. Once, when Wisa landed on my hand to take some food, I was charmed by the crown of perfect flakes that quickly settled on her head, each minute crystal endowed with its own splendid shape. And when the bird flew off, new gems settled on my open hand, until Ked arrived to fan them away with his soft gray wings.

After more than two hours of this aimless, pleasant outing, the jays at last were sated and did not return. I was near the sandbars and I stopped for a time to watch the antics of a pine marten, a relative of the weasel which has considerable tree-climbing abilities. This one, about 2 feet long from tip to tip, was a rich golden brown color and had a heavy, bushy tail about 8 inches long. Like all martens, it was consumed by curiosity. It had heard my approach, but instead of running into hiding, it scampered down from the lower branches of a hemlock and dove into the snow. Poking out its inquisitive, big-eyed head, it inspected me thoroughly. Moving like quicksilver, the alert, attractive animal scampered under and over the snow for several minutes before it tired of watching me. The last I saw of it was the great tail disappearing high up among the evergreen boughs.

It was one o'clock when I got back to the shack and prepared a fancy lunch for myself. Opening a second can of corned beef, I sliced half of the contents, made a heavy batter out of flour, powdered milk, garlic powder, and a few dried herbs, and turned each slice of meat into a sort of primitive Scotch egg, which I fried in the pan. After I had eaten, I decided to make myself a wilderness Christmas cake from a Welsh recipe that reaches back to the Middle Ages. I had all but two of the ingredients called for, but I substituted lard for butter and did without the 1 egg that was needed. The recipe is as follows: 2 cups flour, 2 teaspoons baking soda, 6 ounces butter (lard), $1/2$ cup currants, $1/3$ cup sugar, 1 egg (omitted). The ingredients are mixed with milk (powdered, in my case), and nutmeg is added to taste. The dough is then cut into biscuit shapes about half an inch thick and is cooked on a griddle or, as I did, in a heavy frying pan in which a little flour is first sprinkled to keep the dough from sticking. The pan should not be too hot, to allow the cakes to cook through without burning. After the cakes are done, each is rolled in sugar. The finished product must be eaten fresh, or it becomes stale by the next day, which was of no concern to me, because I fully intended to eat the goodies that evening.

In fact, I planned to have Christmas dinner on the eve of

the festive day, intending to return to my task at first light in the morning. To this effect, I also prepared my main course, a sort of stew made with rice, soy beans, oatmeal, garlic powder, oregano, bay leaves, the remainder of the corned beef, shredded, two OXO cubes, and a package of chicken noodle soup. For an added treat after eating the cake, I was going to make some snow ice cream, a dessert that must be made outside. For this, one lightly packs new snow in a bowl or dish, mixes in some sugar (to taste) and, if available, some vanilla (well-chilled) for flavoring, and enough powdered chocolate to give the dish added zest and color. It is quite good and has the advantage of being much less fattening than the real kind.

For a while during that afternoon I busied myself making a wreath out of pine boughs embellished with poplar-twig sprigs; these I dipped into some wax melted from a red grease pencil. When it was finished I hung it on my door and gathered a number of pine boughs to decorate the interior of my one-room palace. As an afterthought, I found one of my socks that was rather the worse for wear, filled the toe with nuts and raisins and soybeans, and hung it on the wall of the shack next to the door, for Wisa, Ked, and Jak to find in the morning.

By the light of the kerosene lantern, listening to the crackling of my fire, I enjoyed a hearty supper, relished the Welsh cakes, mixed the snow ice cream, and topped myself up with good, strong black coffee and a glass of warm brandy. Afterward, replete, I was about to read when I decided that Christmas could not be properly celebrated without carols, so I stepped outside and sang half a dozen festive songs to the wilderness as well as to myself. That done, I went inside to reread Steinbeck's *Sweet Thursday*, satisfied that none of my neighbors could possibly accuse me of being a Grinch.

I was awakened from a deep sleep by caterwauling of such awesome volume that I at first thought it was coming from inside the small shack. Fumbling for the flashlight as the gargantuan row continued unabated, I realized that what I had at first taken to be one sound was in fact two, each as loud and nightmarish as the other, yet different in tone and pitch, if such musical terms can be applied to the furor to which I was listening. I knew, of course, that the lusty cries were being made by the two pumas, but as I switched on the light and noted that the time was 2:15 A.M., I could not accept what my senses were suggesting—that the cats were almost outside my door!

Setting the light on the floor so that it would not be likely to disturb the amorous duo, I dressed hastily while continuing to listen to the continuing performance.

The female's voice was unmistakable, but her screams, though similar to the samples she had uttered earlier for my benefit, were deeper and of longer duration; they began in low key as a sort of slow moaning that gradually rose to high pitch. The tom, on the other hand, resorted to many growls and snarls, but here again, the quality of these was different from the more usual angry outcries made by the breed. Intermittently he whistled, if that word can actually describe the kind of noises he was making. In each case, the noise volume increased and decreased passionately, the highs becoming incredibly loud, then dwindling suddenly, as though the very intensity of the calls had some muting effect on the vocal cords.

By the time I was dressed, the racket had been going on for about three minutes. It was still in full concert when I slowly opened the door and stepped outside. The moon was in its last quarter and had risen at midnight, but although the snow had stopped falling, the sky remained heavily overcast. The result was that I could not see beyond a few yards in any direction. The caterwauling, however, allowed me to determine the approximate whereabouts of the two pumas, which, contrary to my earlier opinion, were not near the shack, but somewhere on the other side of the collapsed mine building. Knowing the land in that particular area, I suspected they were somewhere in the vicinity of the tom's trailways, at the point where he had made several scratches midway between the end of one major pathway and the start of another one.

Without natural light, I knew there was no hope of gaining a distant view of the pumas. On the other hand, if I switched on the flashlight, I would undoubtedly interrupt the prenuptial duet. These things were evident, but what concerned me more at that moment was the effect that my presence could have on the pair. They might interrupt their amorous duet and run into the forest, or they might resent the intrusion and vent their frustrations on me. I was torn between a consuming desire to witness the rare spectacle that was taking place beyond my sight and a healthy fear of getting mauled, perhaps killed, if the two big predators should decide to react aggressively. Principally, my fear was prompted by the female cat, who had already given me a good scare. The Ghost, I felt, would not molest me.

As I examined my dilemma, the quality of the screams suddenly altered and I heard the thrashing of bodies. Both pumas began to snarl and growl angrily, spitting intermittently; the noise of movement continued.

Instants later the lions became silent; immediately afterward, an amorphous shadow materialized on the south side of the piled timbers. It appeared to be moving toward the trail that led to French Creek. On its invisible heels came another equally indistinct form, but as the two neared the trail, I was able to determine that the larger, leading outline was the Ghost and the somewhat smaller one was the female.

Just before they disappeared, I shone the light on them. The tom didn't turn his head, but continued running, fully stretched out and sending the snow flying in all directions; the female looked my way once before following her mate. Taking advantage of the fact that he was breaking trail through the snow and furnishing her with easier travel, she had almost caught up with him when he entered the creek pathway.

It was clear that neither of the pumas had been aware of my presence outside the shack when they altered their patterns of behavior, but I could only guess at the reason for the change. From the sound, I was led to believe that the Ghost had sought to become overly intimate before the female was ready. She had repulsed him, swatting him with a paw, claws unsheathed, and growling her displeasure. The Ghost had snarled in pain and frustration, but probably continued to press his affections on the female, at which she became properly enraged, charging him. The last thing that a male cat wants at such moments is to fight the object of his affections. Nevertheless, he cannot simply accept her powerful punishment. So the Ghost turned tail, running toward the creek, no doubt knowing full well that the female would follow and that her anger would be replaced by ardor after a good sprint.

In the wild, the females of the species are in full control of the mating rituals. Obeying the demands of body chemistry, they show their willingness for male companionship readily enough, but they do not enter lightly into the act of consummation. Only when they are ready will they accept the male, yet if they see that he is cooling off, they will immediately offer encouragement. Humans would mistakenly call this kind of behavior coquettish. Although it may appear that the female is teasing her suitor, she is, in fact, delaying mating until the time most favorable for fertilization. After the initial mating, and if the first egg is fertilized, the female releases the next egg and is again willing to accept the male. When she stops ovulating, the mating period ends, usually because she discourages the male in no uncertain manner.

That night, too excited to sleep, I made coffee and sat on my bunk, waiting anxiously for the arrival of dawn so that I could look for the two pumas. I thought it likely that the pair would be bedded down somewhere in the vicinity of my base now that they had made contact and initiated the first phase of their complex courting rituals. If luck was with me, I might be able to observe them from nearby.

As I was preparing to leave at five-thirty in the morning, it struck me that my decision to rest the day before had proved fortuitous, although the animals might have decided to court behind the tumbled-down mine building in any event. But it was certain that I would have been forced to abandon my search or observation of them at nightfall because, without moonlight, it would have become impossible to go on.

Following the tracks was easy that morning. The quarter moon, not due to set until noon, was still mostly obscured by the clouds, but the sky had cleared sufficiently to allow a feeble light to permeate the wilderness. Helped also by the reflection from the snow, I was able to walk at a normal pace in the broad trail that the animals had left. After an hour, and about two and a half miles north of my base, at the place where the French Creek valley becomes slightly wider, the double set of tracks presented a new appearance. Until now, the trail showed that the cats had been wading through the snow most of the time, occasionally bounding over places where drifting had occurred. Now, though the pumas had continued to sink into the white cover, they had separated, the female cat's tracks going upslope, the Ghost's angling downward toward the creek, into an area where there were many willows and clumps of tangled rushes. In addition, the tom appeared to have been hopping from place to place, evidently standing on his back paws, raising his forequarters, and leaping lightly about six feet at a time, then repeating the maneuver. The marks left by his passage were so well defined and visible, as first light bathed the valley, that I did not have to leave the pathway that runs slightly above the bottomlands, which until this point we had all been following.

I slowed my pace and walked as silently as possible, interpreting the separation of the pair to mean that they had been hunting or were about to start stalking some prey animal. At most, the pumas were only four hours ahead of me, and this meant that if they were preparing to make a kill, or had already done so, I was in an ideal position to witness at least some part of the action. Nevertheless, bearing in mind the female's behavior on the occasion of our encounter in the vicinity of the Goldstream River, I was keenly alert, scanning her trail with at least as much care as I was devoting to the tracks left by the Ghost. The three flares I carried in the pocket of my parka (I had replaced the used one from the box of six I had brought as part of my supplies), only partly allayed the unease I felt.

The Ghost was predictable. He had given me every reason to trust him; but the female was an unknown quantity who had already shown something of her temper. No doubt the onset of estrus accounted for her display, but that was small comfort, for by now she was fully caught up in the mating urge and might therefore become even more intractable.

Musing in this way, I reached the place where the track I was using dropped down toward the valley, a deviation caused by the sudden rise of land. From here I was given an unobstructed view of the widest part of the narrow valley, a locale covered by tall hemlocks that grew widely apart, giving the land a parklike setting.

Below me I saw both pumas. They had made a kill and were eating it.

Crouching and using the glasses to shorten the two hundred yards that separated us, I saw that they were gorging on a bull moose, a young animal, by the look of it, but quite large. The Ghost was feeding on the uppermost shoulder; the female cat had her head buried inside the abdomen. Since a few tendrils of steam were rising from the cavity she had made, and only a small amount of meat had been taken by the tom, I judged that the moose had been dead for no more than an hour.

Near where I was crouched was a fallen pine, its top plentifully supplied with dead branches that would make an excellent screen. I moved behind this, removed the packsack, and took out my bedding roll, using it as a seat. Settled there, I watched the pumas all day.

The cats fed for just under an hour, gorging themselves to the point where their stomachs became visibly distended. Afterward, each washed meticulously, then they played like two gigantic kittens. The female started the game by approaching the tom and batting him swiftly but lightly on the nose. As the Ghost reared and lifted one of his enormous paws to strike back, the female jumped sideways and rolled over on her back. This was an obvious invitation to romp, and the two spent the next few minutes rolling over and over in the snow, growling in mock anger and struggling mightily. The Ghost was usually on top, but now and then the female wriggled out from under him, turned quickly, and pounced on his back, actually biting his neck. The tom would then roar his displeasure, slide out from underneath his mate, and, as if to tell her that she had been too rough, cuff her two or three times in quick succession. Throughout the furious and seemingly dangerous play, both pumas were careful to keep their claws sheathed; and if the bites that were delivered were obviously lusty enough to cause protests of pain, no blood was drawn. As suddenly as she had begun the game, the female broke it off, turning and striding up the far bank and settling herself at the foot of a big cedar.

The Ghost stretched, yawned, shook his head; sitting, he scratched at an ear, yawned again. He walked to the kill, sniffed at the place where the female had been feeding, licked twice, then turned to face his partner. A moment later he moved toward her, bulldozing snow with his chest as he climbed the slope and stopped beside the recumbent female. Bending toward her, he began to lick her head and face. She visibly enjoyed the attention and occasionally turned her muzzle to lick his chest.

I couldn't hear them across the distance, which had now increased to about three hundred yards, but I was sure they were both purring.

The female yawned and stretched, and as she was doing so, the Ghost jumped for the tree trunk, embraced it, and began climbing, pulling himself up with his forepaws, anchoring with his hind feet, and pulling himself up again, continuing in this way until he reached the midpoint of the tree some fourteen seconds later. The female copied his actions, climbing in much the same way, but seeming to push with her hind feet as she pulled with her front. She climbed more slowly, taking sixteen seconds to join her mate, who by now had draped himself over a couple of stout branches. The female paused long enough to lick his face, then climbed higher, until she too found a comfortable perch. It was 10:15 A.M.

At 2:45 P.M., the female began to descend the tree, but it seemed that the Ghost was not yet ready to stir himself. She reached his perch, stopped, balanced on three branches, lowered her muzzle, and licked his face repeatedly. Unable to resist such blandishments, he perked up remarkably—so much so, in fact, that he led the way down the tree and back to the kill, where they both fed rather halfheartedly, nibbling bits of meat, resting, batting at each other gently with their forepaws. Then, as she had done earlier, the female jumped up, swatted the Ghost's head, and again rolled over on her back. Much as before, they engaged in a long and boisterous romp, but when the Ghost, seizing an opportunity when the female was on all fours, sought to consummate their union then and there, she turned on him in fury, screaming her great banshee wail. As soon as the last high note was launched, spitting and snarling, she crouched and offered her mate the awesome picture of her gaping, fang-filled mouth. The Ghost backed up, but his stance suggested that he was merely waiting for another opportunity for intimacy. This caused the female to retreat also, whereupon the Ghost jumped at her and was met by a savage riposte that gashed the tip of his left ear, causing it to bleed. It was now his turn to growl and snarl, but he took the hint and turned away from the irascible female. At this, she became coy. Jumping up, she went to him and began to lick his muzzle, but when she reached the cut ear, she devoted more attention to it than he apparently enjoyed. Twice he moved his head away from her, but she seemed to relish the taste of his blood and pursued the gored ear. The tom was patient. He actually allowed her to lick the wound several more times before he turned and marched resolutely toward the forest on the east side of French Creek, his powerful shoulders straining as he pushed through the deep snow. The female followed. Soon they disappeared into the trees.

The kill had remained undisturbed by other animals while the pumas were in its vicinity, not even the ubiquitous ravens coming to peck at the carcass, but moments after the predators had gone, nine of the big black scavengers showed up. Cawing and cackling, they swooped on the carcass, one bird actually venturing inside the stomach cavity. Watching them, I pondered my next actions. I was loath to follow the two pumas, not so much because I was concerned about my own safety, but because I did not want to do anything that might cause them to separate without mating, or worse still, leave the vicinity altogether. After a few minutes of silent debate, I decided to stay where I was, watching the kill, not because I expected the pumas to return during daylight, but to observe the remains and to note those animals that might be attracted to it.

The temperature had moderated. It was about twenty-five degrees above zero now, so I was quite comfortable as I scraped snow from the tree trunk at a place where there were sufficient dead branches to act as a screen. Using the sleeping bag as a cushion, I sat down, notebook and pencil in hand, and devoted myself to the vigil.

The ravens continued to feed greedily and even chased away two jays that came to share the meal. But when a fox appeared and followed the trail of the pumas to the carcass, the birds took off for nearby trees and screamed their annoyance. Soon, however, they flew down again and continued to feed, ignoring the fox, which was evidently wise enough to refrain from trying to catch any of them.

For the next hour a succession of small animals came to feed from the remains. The first of these, after the fox had gone, was a weasel, who popped up out of the snow about ten feet from the carcass, sniffed, raised itself for a better view, then dived under the white blanket again, the next time emerging beside the dead moose. It disappeared inside the carcass. Next came deer mice; I counted eleven of them. They appeared one at a time and copied the weasel, traveling underground until near the carcass, then popping up, scurrying on top of the snow, and stationing themselves on top, inside, or below the remains as they nibbled avidly. When a snowshoe hare loped over and began chewing pieces of meat, the mice did not react, but when a lumbering wolverine came racing through the snow with that strange rolling gait typical of those scavenger/hunters, mice, hare, and weasel ran away simultaneously.

The wolverine appeared to be a male and was larger than any of its kind that I seen before. It ate to near bursting, spending thirty-five minutes at the carcass. When it clearly could ingest no more, it climbed on the remains and began spraying musk all over the kill. From a distance of two hundred yards, because the wind was blowing in my direction, I smelled the strong acrid scent of the animal's musk. His action ensured that the pumas would not now return to the kill, which was tainted pretty badly. The wolverine has a simple philosophy in this regard: If it can't carry away the spoils, it sprays them, thus preserving them for its own use or, at worst, to share with one of its fellows.

It was now late afternoon and the sun had disappeared into the west, although at no time during the day had it actually emerged from behind the cloud cover. Satisfied that I had already learned a good deal, I elected to retrace my route home, only then remembering that this was Christmas Day.

THE WEATHER HAD BEEN UNDERGOING THE usual freeze/thaw fluctuations that in the mountains result in floods, slides, and periods of severe icing, a succession of climatic events that began in March with a sudden intense thaw that was immediately followed by a severe drop in temperature. This caused ice to form everywhere and created conditions that threatened the existence of most animals living in the region. It was virtually impossible to walk on any surface; landslides occurred at regular intervals, most of them of a relatively minor nature, but others displacing many hundreds of tons of rocks, trees, vegetation, ice, and snow that slid from the heights with cataclysmic roars. One of these occurred just before noon on March 20, above the place where the Ghost and his mate had killed the moose on Christmas Day. From inside my shack, the dull, prolonged noise sounded like the clangor of an express train running through a tunnel. When the slide was over, it had created a scarred area about one-quarter of a mile long by some six hundred feet wide, the accumulated detritus at its base mounting to almost twenty feet above the grade.

Most predators, including the pumas, went hungry for a period that lasted almost two weeks. The prey species, although able to find local food supplies, had great difficulty in moving to fresh feeding grounds, as a result of which the night raptors became bloated on hares, for these birds, the great horned and the great gray owls, were about the only hunters not greatly influenced by the icy conditions.

For a full week I remained close to home, but on the morning of March 26, after a blizzard had dumped about two feet of heavy snow over the land, I tried a journey up the French Creek valley. This was when I located the slide that I had heard, which had also caused the ground to shake slightly in my immediate neighborhood. Among the tumbled trees and rocks I found the carcass of a large wolverine; it had been caught in the path of the landslide. The ravens led me to the unfortunate animal, which was by that time reduced to a sack of putrefying meat and bones avidly pecked at by six of the large scavenging birds.

Walking through the heavy snow that covered a layer of ice was a difficult business. As a result I spent only two hours in the valley, a time during which I saw no animals and was unable to find any tracks on the surface of the snow.

Back at base, I passed the time reading, making occasional notes, and taking short walks around the valley of the shack, day by day becoming more impatient with the weather as I wondered what the pumas were doing and how they were surviving. I spent almost another week in this fashion, watching anxiously as the weather slowly improved, the sun rising higher and melting more snow and ice. Then, on April 1, a veritable heat wave descended on the region. Overnight the temperature climbed from twenty-eight to seventy-four degrees, the resulting fast thaw causing all the waterways to overflow and race turbulently downslope. The Goldstream tripled its width, carrying cakes of ice, broken trees, and the bodies of fish and animals; French Creek roared down the valley, choked with flotsam, and spread itself over more than half a mile of country in the area of the sandbars, combining with the river to entirely flood the marshlands that lie north of the Goldstream. Now I could move about in relative freedom, but I had to confine myself to the higher land until more seasonal weather came to slow the waters and to allow the Goldstream to dump its overflow into the Columbia.

April 10 marked the advent of true spring in the Selkirks, as far as I could determine. The temperatures were normal for the time of year, a number of early migrant birds had returned, and the Ghost had left his footprints in the mud on both sides of French Creek. I waited one more day to allow the sun to suck up a little extra moisture from the land, and then prepared myself for a trip to the territory of the female cat, whose den I had discovered in late February, almost sixty days after she had sought out the tom and presumably mated with him during the twelve days that the two remained together.

Between Christmas and the last week of February I had devoted myself almost exclusively to studying the Ghost, who had by now become so accustomed to my presence in his territory that he hardly reacted at all whenever we met, or when he knew that I was in his neighborhood and observing him from a respectful distance. Because of this relationship, I was able to watch during two occasions when he stalked and killed mule deer, the first attack taking place in the valley of the Goldstream about four miles east of French Creek. Here, on the north bank, in an area where the land rose abruptly and overlooked a fairly open section, the puma alerted me to the presence of game when he adopted the now familiar stalking posture, holding himself low to the ground, turning so that he was well into the wind, and approaching slowly and carefully, using the trees for cover.

I had caught up with him during midafternoon and now, at early evening, was only about 150 yards behind him. When he first detected the prey, I was unable to see the animals, but when I climbed higher and moved ahead for about 200 yards, I saw seven mule deer does browsing on young aspen that grew in a scattered group near the rising land.

The Ghost climbed also, attaining another hundred feet of height before he began his descent. From my vantage point I noted that he intended to spring on one of the animals from an elevation of about thirty feet, at a place where a ledge traveled northward and afforded a good jumping-off platform. This stalk took twenty-four minutes. Then, as the puma

positioned himself and began bunching his body in preparation for the leap, something alerted the does. They started
to run toward the west, but one, slower than the others,
lagged some distance behind. The Ghost leaped, missed his
target, leaped twice more, and just managed to reach the
deer's hindquarters. He reared, embraced the animal with
both front paws, causing her stride to falter, and then climbed
right on her back. As she was falling under his weight, he
reached for her nose with his left paw, bit into the back of
her neck while anchoring himself with his three other paws,
and then wrenched the deer's head violently to the left. The
action was well coordinated and swift, so that it seemed as
if all movements were taking place at the same time. The
deer's neck snapped even as she was falling to the ground.
She was dead instantly.

Emboldened by the tom's acceptance of my presence, I
walked down to the valley and stationed myself at a place
about 100 feet from the kill. The Ghost was already eating
when I arrived, but apart from lifting his head to give me a
cursory inspection, he remained undisturbed. On this occasion he did not disembowel the animal, but began to feed
on her right haunch, taking from it an estimated twelve to
fourteen pounds of meat during a period that lasted thirty-
two minutes. After he had washed fastidiously, he strolled
away and climbed a tree about 250 feet west of the kill. This
successful attack took place on January 7. The second kill I
saw was made on January 16, also in the Goldstream valley,
but in the area of the swampland. This time the puma broke
the deer's neck by striking the animal on the shoulder with
both front paws, much as he had done to the caribou.

I did not see the lion make any other kills after that day.
But I was able to learn that in every case the deer, moose,
or caribou attacked received a broken neck, which suggested
that this particular method of killing is probably employed

the most frequently. Well executed, such a technique quickly disposes of the prey and lessens the lion's chances of being injured by a struggling animal.

All the members of the deer family will kick in self-defense, and because their hoofs are pointed and extremely hard, these can be fearsome weapons. In a confrontation, moose, caribou, elk, and deer stab at an enemy with their forefeet, vicious blows that can easily kill a wolf if they connect. A puma, being larger and somewhat better protected by a thick hide, might escape such an attack alive, but it would certainly receive wounds, any of which would impair its efficiency. A lion's chances of survival after being injured in this way are minimal, for it would most likely starve to death.

January passed swiftly for me that year, but it was a month during which I saw more of the wilderness at night than during the day, spending a total of nineteen nights tracking and observing the puma from one end of his range to the other. I discovered with the arrival of February that my human, day-oriented senses had become exceptionally well attuned to the nighttime environment. Before this, I had believed that I was familiar with, and proficient in, the world of darkness. But I had never before exposed myself to such frequent and lengthy sojourns within the darkened forests.

I did not at first dwell consciously on this aspect of my work, but after the fourth month of my stay in the Selkirks, I realized that my hearing had been brought to much finer pitch and had become capable of picking up and recognizing even the lowest sounds. My sense of smell, which had always been good, became acute, while peripheral vision improved to the point where I could move through the night wilderness almost as quickly as I could during the day. But it was the improvement of hearing and smell that most surprised and pleased me. These two senses, which are woefully neglected by civilized humans, were now allowed to function in conjunction with, and as an adjunct to vision, and since sight under such conditions was the least efficient of the three, I began to place subconscious reliance on ears and nose, especially the latter. I habitually smelled the immediate environment, often getting close to the ground or to objects of interest and sniffing them intently. In this way I expanded my perceptions of odors, discovering many new ones and storing them in whatever portion of the memory bank catalogues and retains the smell of familiar objects. I learned to identify the lion's scent, as well as those of bears, wolves, and a number of other animals to the extent that I could often pick them up from plants and trees, dried grasses, urine, feces, and the rocks that the animals had rubbed against. When near enough to some animals, I could even detect their general location by smelling the air. If the wind was favorable, I could scent the puma from a distance of about fifty yards, and on two occasions before snow time, I detected fresh lion scratches while I was still one hundred yards from them.

During February, after becoming consciously aware of the changes that affected my hearing and sense of smell, I began to pay particular attention to them as I continued my day-to-day studies of the pumas and the world in which they lived. Of particular interest to me was the realization that humans inhabit what is virtually an odor-deprived environment, a circumstance brought about by physical changes resulting from evolutionary modification as well as by the fact that for thousands of years humans have lived in tamed, synthetic environments. As the centuries passed, a stage was reached when personal security was attained by mechanical means, the availability of sophisticated weaponry, and the constraints of civilization. As individuals, humans were no longer hunted by animals; neither did they need to hunt in order to survive. By this time, mankind had actually altered the environment and entered into an unnatural form of existence without realizing it.

In recent times, and especially since the start of the Industrial Revolution and the technology that this produced, science has been able to measure the capacity of some human and animal senses. We have learned about hearing and vision, appreciation and perception of color, the mechanics of taste, and even some of the mechanical aspects of smell, but to date we have been singularly unable to really *measure* odors and classify them. This is demonstrated by the fact that whereas we have names for all primary colors and most, if not all, of the shades that mixing these produce, we must still rely upon a description of odor by likening each smell to the aroma of some object that is universally familiar: the smell of fresh-baked bread, the fragrance of a rose; the odor of putrefaction, and so on. Beyond this, humans have arbitrarily divided scents into only two classes: good and bad. In the wild, smells are smells, neither good nor bad but important for the information that they contain. In like manner, most humans note only exceptionally strong odors and fail to detect slight nuances of aroma.

Unadulterated by perfume, scented soap, or deodorant, every human produces his or her own highly individual odor. So do animals, but whereas they recognize each other by scent rather than by sight, most humans usually fail to detect, or are sickened by, the body odor of other individuals, with perhaps the exception of the smell of those to whom they closely relate—and even this is but casually noticed, recognized at the moment of olfaction and usually forgotten when contact is broken.

Awareness of these things and my own heightened odor perception exposed me to a whole new environment and allowed me to relate more fully to the pumas as well as to all the other animals whose world I was then sharing. Sometimes I would go so far as to crawl along a trail, my nose an inch above a track; and my satisfaction was great when I could isolate the smell of the animal that had made the mark. This was not always possible, but it happened enough times to teach me that humans could, indeed, track by scent.* Each time I succeeded in this task, I congratulated myself and felt somewhat superior until I thought of a wolf, or of the Ghost, who could follow a scent so accurately that it was as though they were tied to the odor by an invisible thread. And I recalled seeing many animals, especially predators, who tracked by scent and ran casually along the trail with their noses one or two feet above it and facing in the direction of travel, yet were able to pick up the odor of each track as they ran. Such feats were quite beyond my capabilities; I felt humble and at the same time envious of the abilities of even the most insignificant animals of the wilderness. A tiny deer mouse, scurrying along under the snow, could actually scent little mounds of trail rations that I had left for the jays, guiding itself unerringly to the pile and popping up suddenly an inch away from the supply; a chipmunk can smell a peanut in its shell from a measured distance of ten feet, yet when I try to scent the nut, I can only barely do so by actually pressing it against my nostrils.

Principally, I believe, the absence of civilization helped to a great extent to develop my senses of hearing and smell.

---

*In the upper northwest corner of Brazil, deep in the jungle region of the Içana River, live (if they have not now become extinct) a tribe of Indians, the Cihuma, who hunt by scent, having an almost uncanny ability to smell objects from considerable distances. Evolution also modified the faces of these people, shortening and decreasing the length of the nasal passages, just as our own muzzles and nostrils have been altered; yet the Cihuma did not lose the keen sense of smell found now only in them and in wild animals.

The only "civilized" odors to influence me were my own, the great majority of which were natural, even to the smoke of my fire. No longer was I being subjected to the harsh and poisonous fumes of gasoline, diesel fuel, and the hundreds of other alien smells that hang over urban centers like some great, smothering miasma. Instead I was enveloped in the odors of the wild, the scent of evergreens, clean air, the musk of animals, fresh mud, mulching vegetation, fungi on trees or in the ground, dead needles and leaves, my own odors and those of my equipment, such as leather, canvas, food, and many more.

Similarly, my ears were not assailed by thundering trucks and poorly tuned automobiles, the crashings of construction, the cacophony of crowds, the warlike roar of aircraft engines, and all those other sonic disturbances that characterize any modern civilization. In their place were the voices of the animals, the songs of the birds, the gentle breeze through the crowns of trees, the primordial howling of the wind during storm conditions, my own breathing, and, on those occasions when I spoke, the quiet sound of my words, each of which emerged singly and meaningfully. Cracking twigs were warnings, telling of movement made by something somewhere, the volume of sound they produced offering clues that often identified the cause of the disturbance before the eyes or nostrils became aware of it. A rolling stone, small or large, had meaning, as did any almost imperceptible rustle. Every single noise encountered in the wilderness was a signpost to some form of activity, even if it was merely caused by gravity, when a twig might fall from a tree, or a pebble roll down a gradient. My world of sound was full of signals, and all of them had meaning once I learned to find it.

My sense of touch also developed. I learned the braille of the wild, using the tips of my fingers to trace the outlines of a track or path in the blackness of night or to guide me through obstructions or merely to feel objects for the pleasure that this exercise gave me.

Because odor and hearing had become so important to me, I even refined my movement. I had learned years earlier to walk quietly through the wilderness, but now, eager to absorb every scent, sound, or sight offered to me, I controlled my movements to the point where I could travel almost as silently as the animals, although my two legs could never attain the utter stealth of the four-footed walkers.

On February 26, I set out at first light to look for the den of the female puma, convinced that by now, with two-thirds of her gestation time already passed, she would have found for herself some secure cave in which to deliver her kittens. I searched all morning under blue, sun-filled skies, realizing as I was doing so that I had added one more accomplishment to my list, although it was a kind that practically defies explanation: I had developed the ability almost to *think* like a lion, to recognize instantly those places that would not offer prospects of prey, or that were too tangled for quiet travel; I could "read" tracks and tell from the width of the stride and the depth of the impressions whether a cat was walking unhurriedly—alert, but not stalking—or whether it was intent on hunting, but not yet within sight, scent, or sound of game.

Of perhaps greater importance that day, I could also tell when a puma was undertaking a purposeful journey; if it was, in other words, principally concerned with going from A to B and only casually alert to its environment. In this case, the distance between strides only varied by an inch or two and the course was direct, involving many shortcuts and leaps over downed trees or small rocks so as to shorten the distance while always aiming in one particular direction, no matter what detours had to be made because of terrain. Actual stops during such journeys were rare, and exploratory side trips were never made.

By two o'clock in the afternoon, having followed just such a spoor uphill for more than two hours, I at last found the den. It was unoccupied at the time and located above the three-thousand-foot level on the westernmost slope of Downie Mountain in an area immediately above Goldstream Creek. Some ancient upheaval, or perhaps a prehistoric slide, had created a jumble of rocks at a place where the grade ran steeply and where, because of the exposed granite, few trees had managed to take root.

The blocks, slabs, and boulders were scattered over a section of land that probably measured one hundred yards in length by about fifty in depth, but was of irregular formation. In the middle of this jumble of stones was a huge rectangular boulder, lichen-encrusted and lead-gray, that rested right against the cliff face and was partly covered by three big granite slabs that formed a grotto, the entrance to which was two feet wide by somewhat less than that in height, an opening resembling the embrasure of a defensive pillbox.

The cat's tracks led me to the den area, below which I stopped to examine the fortresslike rocks through the field glasses, not wishing to force a confrontation if the puma was inside. The den was easily seen from my downslope position, as was the beaten pathway that the lion had made during her comings and goings.

I was about seventy yards away, near enough to throw a stone so that it would land among the lower rocks with a rattle that would certainly cause the lion to investigate if she was at home. The missile, about the size of a golf ball, landed squarely on one rock, made a modest noise, and fell into a patch of snow. I waited some moments, but when nothing happened, I rummaged under the snow until I found a larger stone. This one hit the top of a leaning slab, making a good crack; then it rolled downhill with a satisfying loud clatter. No cat.

Confident that the puma was not in the den, I climbed toward it, but I whistled a loud tune as I did so, just in case. When only a few yards from the entrance, I picked up another rock and threw it inside, listening to the noise that it made. Then I bent down, preparatory to crawling in. Now I smelled her odor, mixed with the slightly rancid smell of frozen meat.

Inside, the den was somewhat oval, running about six feet toward the rock face; it was about nine feet wide. The roof sloped from ground level to a height of about four feet where the slabs rested against the heavy boulder. Three feet from the entrance were the remains of a mountain goat, although little was left of the animal except bones, pieces of skin, clumps of white fur, and part of the head with the black horns still attached. One shriveled eye remained in its socket, staring at me out of the shadows; the other had been ripped out with a piece of the skull, which was bare to the bone and showed the marks of the cougar's teeth.

To the puma, that goat had been much-needed food, but to me it was as grisly a find as any I have ever made, principally because of its funereal location and because I was lying flat on my stomach and practically eye-to-eye with the ghastly, glaring orb. Only then did it occur to me that I would find myself in a particularly nasty position if the cat happened to return at that moment.

I crawled out of the cave, stood up quickly, searched the surrounding wilderness, and beat a quick retreat, greatly relieved that I was quite alone on the mountainside. In my haste to quit the area, I inadvertently took a wrong turn, following another of the lion's trails that led away from the track I had used on the way up. When I realized that my own footprints were absent, I turned at an angle to rejoin the right trail, and in doing so I saw that the land to the north descended precipitously for about three hundred yards before ending as a scree-covered gorge against a vertical wall of rock that was the southern base of a tall, absolutely treeless dog-tooth of crumbling stone. Through the center of this gorge ran a small serpentine creek, now frozen, that dived down-slope after about two hundred yards. The waterway appeared to be taking a straight course toward the Goldstream River, following the steepest face of the mountain in this area, a part of the country that I had not as yet explored. Removing the map from the packsack, I oriented it and found that I had guessed correctly and that the nameless creek emptied into the river not far from the owl pond. I would, I decided, investigate this section of land after the snows had melted.

As the weather began to moderate, clear skies, brilliant sun-shine, and hardly any wind caused a slow melting during the day and ice crusts at night that by morning were strong enough to support my weight, but became crumbly and treacherous by early afternoon. It was hard to get about, especially on snowshoes. I took to spending my time close to home, keeping an eye out for the Ghost but failing to see him, though his tracks were plentiful along the trails near my base. Since the weather was ideal for hunting, I felt sure he was enjoying one of those periodic feasts that result when a predator can run easily on a snow crust that will not support the prey.

I had heard the wolves howling repeatedly since mid-January, their voices coming from the river flats in the general area of the pond, but the pack had not entered the French Creek valley since early December. They too, I felt sure, were hunting successfully right now and were probably keep-ing clear of our valley because the tom obviously considered it his special preserve.

Having lots of time on my hands every afternoon and evening, I spent most of it making notes, reading, and feeling great sympathy for the prey animals, but knowing that soon, with the coming of spring, the odds would be against the hunters and in favor of the hunted. Of particular interest to me then was the charting of the Ghost's kills, nineteen of which I had encountered since my arrival in the Selkirks. I could not, of course, know the exact number of animals brought down by the puma during the last five months, but I did not think that he had taken more than an additional half-dozen of any large species, or I would probably have found more remains during my wide traveling through his range.

The figures I had gathered to date appeared to confirm the supposition that a puma kills between thirty-five and one hundred big-game animals during the course of one year. At this stage of the study, I felt that thirty-five might be a low figure and one hundred a high one, and that somewhere between these numbers might lie the true answer to a lion's annual consumption of meat.

Because of the restrictions imposed by the weather, time passed slowly, and for me this was somewhat depressing. I had been away from my own kind for half a year, and although I was not yearning for human companionship, there were days when I would really have liked to go out and have a sumptuous dinner in some luxury restaurant. Even a fresh trout would have been a rare treat at about that time, for I was getting rather tired of the uninteresting food I was consuming. It was certainly sustaining and had kept me in top condition, but I was quite ready for a change; my memory of the Goldstream River trout caused me to salivate more than once while I was shack-bound. And then, at last, it was spring.

Anxious to become active and especially eager to find out if the cat had given birth to her litter in the mountainside den, I left home at first light on the morning of April 11, my pack containing enough food to last a week and the envelope with its three flares resting in the inside pocket of my light parka, just in case.

Near the pond, as I was approaching the valley of Goldstream Creek, I caught a glimpse of the wolf pack, led by White Spot, as the animals trotted across an opening and then disappeared among the trees. Half an hour later the wolves howled, a series of uncoordinated calls that lasted only a few minutes. It seemed that, as for most of the animals in the wilderness, the advent of spring was causing the pack to become excited, a feeling I shared as I walked in brilliant sunshine and listened to the constant tintinnabulation of the melting snow and ice.

Despite wet legs—in order to reach the island where I had positioned the crossing logs, I had been forced to wade through frigid water-thaw that flooded the lowlands—I felt great contentment and a sense of oneness with land that was so boisterously coming awake after the winter's lethargy. Varied thrushes had returned in considerable numbers, and their wonderful melodies accompanied me as I walked; white-crowned, northern tree sparrows had also returned, and their voices enhanced the sylvan medley while the fast staccato drumming of ruffed grouse cocks supplied percussion. As I was climbing toward the puma's den, at an altitude of about fifteen hundred feet, a ptarmigan cock set up a vigorous clucking and cooing, but stayed hidden within the underbrush, perhaps occupied with wooing a sleek hen.

When I estimated that I was about half a mile away from the rockslide, I angled northward in order to bypass the den, get above it, and find an observation point where I could remain concealed without arousing the puma's suspicions. The detour took me near the gorge I had noted in February, so I paused to have a look at it.

The land was heavily treed in this region, the only open location being the slope of the gulch and the almost vertical mountain cliff on the other side. The bottom, a strip of uneven land about 150 yards wide through which the creek flowed, was sparsely covered by bush growth, but the little stream I had seen in winter was now a vigorous, noisy watercourse that appeared to have risen six or seven feet. Like a ribbon of silver, it ran down the mountain, carrying with it debris, but it was evidently unsilted as yet, no doubt because the topsoil was still frozen in higher elevations. From what I could see of the country to the west, it would be possible to climb down and rejoin the Goldstream valley at a point not far from the old wooden bridge I had found. This might save some time and allow me to explore a section of land that I had not as yet visited. So I decided that on my return I would make the attempt.

By early afternoon I found a place some two hundred yards to the east of the den, overlooking its entrance and about a hundred yards above it. Here, amid another, smaller jumble of rocks, I would be well concealed; a shallow little cave formed by the granite gave me a good shelter in which to spread my bedding.

Having found a refuge, I next used the glasses to examine the area around the den, looking for signs of use. I was elated when I noticed a profusion of small pugmarks in a slow-melting snowbank almost immediately in front of the den. Without a doubt, there were kittens in the lair! Some of the mother's large tracks testified to her presence as well. Half an hour after confirming that the den was in use and that the cat had given birth to at least two kittens, I had set up camp and made myself as comfortable as possible in preparation for a vigil that I hoped would yield good results. The now-empty packsack I folded and used as a cushion on a rock that I positioned near the place from which I intended to keep watch; my food, which consisted of trail rations, pre-cooked bannock bread, two cans of corned beef, and some dried fruits, was near at hand. When I added a few flat rocks on top of the boulders that stood in front of my redoubt, I made a small peephole out of which I could observe the den while remaining concealed from its occupants. My only problem was going to be drinking water. I had brought a three-quart canteen, which was full, but I would need to ration my intake in order to have enough to last for the intended duration of the stay, because now that I was in position, I was not going to be able to move very far from the place.

After an uneventful afternoon, most of which was spent watching and listening to birds, I started eating trail rations as the sun was disappearing behind the peaks. I was still chewing the first mouthful when the cat appeared suddenly outside the den. Distracted with the business of eating, I had missed seeing her crawl out, but there she stood, staring into the distance, head held high, nostrils flared as she sniffed for messages, her ears equally attentive. I activated the stopwatch.

For one minute and twenty seconds she stood like a statue, feet firmly planted, long tail hanging limply, its black-tipped end curving upward; only her flanks moved, a reflexive motion that kept pace with her breathing. Next she turned her head to the right, toward the north, and examined the region for thirty-two seconds before swinging toward the south and checking there for twenty-eight seconds. Evidently satisfied by her careful examination of those three frontal quadrants, but not once having turned to examine the area upslope (for reasons unknown to me), she lowered her head, half-turned her forequarter until she was peering into the den mouth, and made some kind of vocal signal that was too low for me to hear, but which caused her mouth to open and her throat to quiver. Within seconds, three little pumas tumbled out of their lair, the speed of their arrival suggesting that they must have been waiting for their mother's call almost at the entrance. I don't normally like the word *cute*, but to describe those cougar kittens any other way would be inaccurate. They were cute!

Two were about the same size, and the third was considerably smaller. I could not determine their sexes, but I surmised that the bigger ones were toms and the finer, smaller one was a female. They were about the size of housecats, the background color of their coats being a light, creamy fawn on top of which were superimposed many brown-to-black, irregularly shaped markings that became raccoonlike rings around the short, somewhat pointed tails. I estimated that each weighed between 3 and 4 pounds and that they were now at least two weeks old, probably closer to three. With the exception of the black that lines the ears of adults, the little pumas were quite unlike their parents, even lacking the typical high-hipped, low-shouldered stance. But if their length equaled that of a domestic cat, they stood taller and their paws were at least twice as large.

At first, emerging on still-rubbery legs, they stood bunched, staring at the enormous new world that was spread before them like some gigantic, untidy tablecloth, their soft round eyes blinking against the light. The mother, clearly enraptured with her young, leaned over them, licking first one and then the others, an action that caused each little cat to stagger as the great rough tongue slurped over them. After a few moments the mother cat stretched out, her body blocking the den entrance, though whether this was done accidentally or by design, I cannot say. Almost immediately the smallest kitten staggered up to her and began to suck on a milk-filled teat, and was soon followed by the other two. The feeding lasted three minutes and twenty-three seconds for the first kitten and a few seconds longer for the other two. Then the mother licked each one thoroughly, holding her unwilling offspring with one enormous but gentle paw when the little ones sought to escape the attention.

When the toilet was completed to the mother's satisfaction, each kitten was allowed to wander at will, something they did with great concentration but some difficulty, for their legs, and especially the big paws, didn't seem able to coordinate movement. Now and then, one or the other would take a tumble, and lie struggling for some moments before regaining its feet. One of the larger kittens tried to scratch its neck with a back paw, but it immediately fell over and rolled a little way downslope, letting out a plaintive, mewling cry that reached me faintly. The big cat ignored all this; she lay relaxed, staring into space, now and then glancing at her young.

Despite their poor coordination, it was not long before the two bigger kittens began to wrestle, and as they tumbled awkwardly around the den "terrace," the smaller one joined in; for some moments there was a confusing, writhing mass of dark-spotted fur from which a ringed tail occasionally protruded. While this mock combat was still going on, the mother rose, stretched and yawned, opened her mouth, and evidently uttered another unheard call, then turned, crouched, and crawled into the den. One by one the kittens followed, led by the smallest. One of the two bigger cubs became momentarily distracted by a piece of dead branch, paused to examine it, then realized that it was alone. Mewling, it sought to run into the den, stumbled and fell, rose again in a rather untidy manner, and disappeared. It was now early evening.

Puma kittens come into the world covered by their woolly, spotted coats and with their eyes closed, which remain so for between eight and eleven days. Like all feline young, they require a good deal of attention from the mother at this time.

After the kittens open their eyes, they become active, but their movements are slow and wobbly. This awkward stage does not last long. By the time they are two to three weeks old they are already playing outside, and continue to live exclusively on a diet of their mother's milk until they are about six weeks old. Then they begin to nibble on meat and bones, nursing less frequently and finally becoming weaned when they are three months old and begin to accompany their mother on her hunting travels, although she usually makes them stay in one place when she picks up a strong scent. If she succeeds in stalking and killing prey, she feeds first, then goes back for her young and leads them to the food, allowing them to eat until they are sated, in all probability indulging in a second meal herself.

Gradually, as the young cats gain size, strength, and experience, they go hunting with mother, and by the time they start losing their spots at six months, they are already quite proficient stalkers and killers of smaller game. But the mother keeps them with her until they are about two years old, when they are almost as large as a full-grown adult and more than capable of earning their own living.

Usually the young puma siblings remain together for at least another winter. Then they split up and go their own ways, strangers to each other if they should meet by chance in the future. Young females reach sexual maturity at thirty months, and toms mature when they are three years old, which is nature's way of lessening the chances of brother-and-sister matings that could lead to a weakening of the species.

For three days I remained quite inactive as I watched the den and was periodically rewarded by seeing the kittens and their mother. I noted that the little pumas were making remarkable progress. Soon after dawn on the second day, the mother returned to her young carrying the body of a ruffed grouse, an offering that she dropped outside the den entrance before making her usual call. The trio of romping cats emerged together, jostling each other in their anxiety to greet their mother and to get first place at the two back teats, which are always better supplied with milk then the four front ones. But the female would not let them feed. Instead she pawed at the grouse, lifted it with her extruded talons, and tossed it into the air. When it landed a few feet away, the sudden thump of its fall and its unexpected appearance in a different location panicked the kittens and caused them to scramble back into the den. Their mother called them outside again and once more encouraged them to attack the dead bird. For a while each tiny puma stalked the inert form with more fear than patience, then one of the larger kittens pounced on it, causing some feathers to fly off; now the others rushed in and a free-for-all ensued, each trying to steal the entire bird amid much juvenile snarling and growling.

The youngsters did not eat any of the grouse, but they forgot their hunger for milk as they played with the corpse, which was evidently what their mother had intended. She now sprawled nearby, lying on her side, but keeping her shoulders and head upright so that she could watch the antics of her infants, the love for them shining in her eyes to such an extent that, recalling the occasion when she had charged, it was difficult to believe that this was the same animal, or even that she could ever be capable of such furious behavior.

The kittens played with the unfortunate grouse for less than fifteen minutes; then, when the smallest turned away and went to the mother and was allowed to suckle, the other two interrupted their game and hurried to get their share. Looking at the grouse through the glasses, I saw that it more closely resembled a bundle of plucked feathers than a bird, yet, apart from a few tears in the skin made by the needlelike claws of the little cats, no part of it had been eaten or even chewed, as far as I could determine.

Later, after the young had satisfied their appetites, the mother stretched out fully and went to sleep. One of the larger kittens snuggled up against her belly and snoozed also, but the small cat and her sibling amused themselves for almost half an hour worrying the grouse and finally succeeding in dragging it some distance from the den entrance. At this juncture a bald eagle, the first I had seen that spring, sailed over the mountain and uttered a series of shrill, high-pitched cries. The mother puma immediately sat up and the kittens scurried into the den. The eagle sailed on, but the kittens did not reappear for the remainder of the day. At dusk the puma left her young, but whether she was going hunting or was intending to visit her last kill and feed from it, I did not know.

An hour after the cat had gone, in full darkness, I got up from my seat and did some silent calisthenics to restore circulation and loosen my stiff muscles before eating my evening meal.

On April 14, with rain falling in a steady stream, I reluctantly decided to abandon my observations of the kittens and their mother. This resolution was not taken because of the weather, but because the female had become intensely nervous and alert and had not left her kittens since the night of the twelfth. Assuredly she was alert to my presence, otherwise I feel certain that she would have come up to investigate from close quarters. Nevertheless, she exhibited intense restlessness and had begun spending fewer hours in the den, while actively discouraging the kittens from spending time outside.

When first light arrived on the last day of my vigil, I knew that the puma had not eaten for some thirty hours and I was worried that she might have abandoned the kittens or, almost as bad, moved them to a makeshift lair elsewhere. The only thing to do was to interrupt the study. I had, after all, confirmed that she had mated with the Ghost and had produced three healthy young that I had been able to watch undisturbed for a total of eleven hours and forty-seven minutes on nine different occasions. I had learned a great deal from this venture; that would have to do, because on no account did I wish to disturb the animals further.

Having arrived at this decision, and while the mother cat was pacing restlessly outside her den, I packed up my equipment as quickly as possible and stalked away from the locale, climbing up the mountain for nearly half a mile and then making a wide detour calculated to bring me back toward the gorge that I intended to explore. When I felt it was time to reverse myself and begin to descend the slopes, I was south of the den, but above it at about the four-thousand-foot level in an area that sloped fairly steeply, but which was well treed and offered reasonably good footing, although the ground was sponge-wet and quite slippery.

As I descended, I turned slightly toward the north every time I took two hundred strides; I wanted to aim myself in the desired direction gradually, so as not to run the risk of disturbing the puma.

It took me until noon to reach the gorge area, but by then the rain had stopped and the sun was trying to penetrate the clouds, so I could find no good reason to abandon exploration of the canyon. The descent was steep, but after studying it for the third time I concluded that it did not appear to present insurmountable obstacles, and the trees that lined the first two-thirds of the sharp slope offered themselves as handholds in those places where scree and mud could pose problems. Before starting the journey, however, I sat down and rested for an hour, eating trail rations and having a good drink of water, the first really thirst-quenching swallows that I had taken since my arrival four days earlier.

Soon after one o'clock I began the descent, finding that the route to the bottom of the gorge was relatively easy. In slightly more than an hour I stood beside the fast-flowing creek, refilled my canteen from its icy waters, rested for ten minutes, and then followed the waterway to the place where it took a dive as it fell over an extremely steep part of the mountain, a place that looked a lot more difficult than the earlier part of my descent. Nevertheless, because of the trees, I felt I could manage this slope as well, but I was somewhat concerned by the fact that I could not now see the kind of terrain that lay ahead, for this was entirely hidden by the pines that grew thickly all the way down the mountain.

Not long after I committed myself to the descent, the rain began again, causing treacherous conditions underfoot, but by then it was too late; it was easier to keep going down than to try to climb back up. I slowed my pace, traveled from tree to tree, paused beside each one, then went down some more, spending another hour in this fashion and finding myself in an area so steep and so packed with reluctantly melting snow that it was unsafe to tackle. In an effort to detour so as to circumnavigate this section, I went from bad to worse, entering an even steeper part where the trees were widespread. Here the sun had melted the lingering snow, but the forest floor consisted of a mixture of mud, silt, and rotting pine needles.

Directly in front of the place where I stopped to survey the land was a long, treacherous slope that was almost entirely covered in scree, at the bottom of which rose a sheer cliff that was shaped somewhat like a giant kidney dish and formed a cul-de-sac. Anything that went down there could only get out again by climbing the scree area!

Such dead ends are not unusual in mountain country and

are always to be avoided by sensible travelers. And that, I thought, as I started to turn around, was exactly what I was going to do. But my right foot slipped from under me; the tree that I reached for to steady myself was just out of my grasp. I fell; and I began to roll, dislodging pieces of scree and heading straight for the granite buttress.

Recollections of that long, bruising slide remain vague. At some point in my descent my head struck a hard object; I remember feeling pain, then nothing until almost four hours later, when the rain brought me back to consciousness. It was dusk. I ached all over, but the pain inside my head was excruciating. Feeling the most tender part, the right side of my forehead near the hairline, my fingers encountered a large bump and the stickiness of blood. General weakness made me realize that I had lost quite a lot of blood. My parka was soaked through with it, right down to my shirt; and despite the drizzle that was falling, the ground was red where my head had been resting. I was still bleeding, but from the feel of the wound it seemed that coagulation had begun to take place.

My hat was lost and the canteen of water was badly dented, but at least I had retained the backpack and, more important, the first-aid kit. Before moving, I tried my legs. Nothing was broken, but there were many aches and the feel of scraped skin rubbing against my pants. My left arm was under my body; I rolled slightly, expecting to feel the knife-edge of broken ribs, but was happy when I was rewarded by dull aches instead.

Now I tried moving my left arm. It worked and didn't even hurt, except for a couple of minor cuts on the back of the hand. Having established that I was still almost entirely whole, I began the business of removing the pack, an experience that I never wish to have to repeat. Coaxing myself to a kneeling position and holding on to the bole of a pine that I realized I had been resting against, I wriggled one arm out of its strap, doing my best to ignore the many acute pains that the movement elicited from various parts of my anatomy. The rest was easy. Maintaining a hold on the tree, I shucked the pack off my left shoulder and held on to it while I rested and tried to ease some of my aching parts.

The first-aid box contained sterile, absorbent pads, cotton batting, bandages, and a bottle of a hundred tablets of a strong, codeine-based pain remedy. Before attending to my bleeding head, I swallowed four tablets, then I applied two pads to the wound, placed a wad of cotton over them, and bandaged my head. By this time it was almost dark, so there was nothing to do but remain anchored to the blessed pine that, although it was responsible for my side and chest turning a deep purple and angry red later on, had at least arrested my descent after some two hundred yards of rolling.

The rain continued to fall. It was now cold on the mountain, and I was very wet. From the pack I took the large sheet of polyethylene plastic that I always carry and draped it over myself before taking off all my clothes, garment by garment, and then changing into dry clothes. This took more than half an hour, but at the end I was at least dry and somewhat warmer than I had been when I began. Huddled under the plastic and leaning against the tree, I began to eat, forcing myself to do so because I felt nausea but not a trace of hunger. Yet I had to weather the night where I was, for the climb back up was too treacherous to undertake in darkness, and in any event I didn't have the energy even to contemplate, let alone try it. I needed nourishment. I had one can of corned beef left and I ate it all. There was also one foil envelope of chicken noodle soup; I ate that also, as it was, dry, but between swigs of water. The stuff was salty and pasty, but it was sustaining. By the time I was ready to start on some dried fruit as dessert, my stomach rebelled; one more mouthful, I knew, and I would be sick.

I tried to sleep, but not before I had lashed myself to the tree with the nylon line I also make a habit of carrying in the mountains. The best I was able to manage during the entire long night were little half-hour naps, but, aided by several more doses of codeine—two at a time now—I managed to control the pain and to allow my body to rest to some extent.

No dawn ever looked more beautiful than the one that arrived in the Selkirks on April 15, 1973!

Drinking water steadily, for I was very thirsty now, and forcing myself to eat trail rations and dried fruit, I waited for daylight, relieved that the rain had stopped and that the skies were clear. When the sun began to show golden in the east, I gathered my things, struggled into the pack, coiled the rope around my neck, and started the treacherous journey up the thrice-cursed slope of scree.

It took four agonizing hours to reach the top, a feat that twice caused me to lose consciousness while I was resting against the boles of trees. Had it not been for the rope, I am quite sure my bones would still be there, by now picked clean by birds and perhaps mice, those being the only things that could possibly negotiate that descent, or would want to! But I used the rope, which I had knotted around my waist before I crawled over the first hundred yards of scree, to fasten myself to a tree whenever weakness threatened to send me for another tumble.

Near dusk I reached the valley of the mine and my humble, but to my eyes wonderful little shack. There, after washing in warm water, I was at least able to inspect the cut on my head. But first, and contrary to medical advice, I downed a cupful of brandy, neat. As soon as I removed the grubby dressing, the wound began to bleed heavily. It definitely needed stitches. It looked as though I had grown an extra set of lips and that these had been painted with carmine lipstick. Aided by the mirror, which is one of those that magnify on one side, and using fine sewing thread that had, together with the suture needle, been boiled for fifteen minutes, I "scrubbed up," trimmed back some hair, and, feeling decidedly lightheaded from fatigue, brandy, and loss of blood, I put five clumsy stitches in the wound, closing it quite well, but causing something of a pucker. Afterward I slept for sixteen hours.

MY ACCIDENT KEPT ME PRACTICALLY HOUSE-bound for the entire second half of April, a time when I was incapable of performing activities that were not dictated by personal survival. The first four days were the worst. Although I rested a lot, food was abhorrent and I had to force myself to get up in order to perform those tasks that were necessary to my well-being, such as going out to get firewood and attending to the demands of my body.

I did not require a medical degree to diagnose my condition: I was suffering from concussion. In addition to those occasions when I had lost consciousness, I was pale under my suntan, my pulse was weak, my skin was constantly cold, and I became nauseated at frequent intervals. My self-prescribed treatment consisted of bedrest, fluids, some soup when I could get it down, OXO broth, milk, and codeine tablets, but my headaches were so severe during the first ninety-six hours that I could only manage to sleep for a few hours at a time; on the other hand, when I was forced to move, I was almost instantly dizzy and mentally sluggish.

On the fifth day my headaches became less severe, and this was such a relief that I hardly took notice of the pain that followed every movement of my badly bruised body; I had felt this all along, of course, but it was of minor concern compared with the head pain. The forehead wound, however, was healing nicely and showed no signs of infection when I removed the stitches.

Once I was satisfied that I was not suffering from severe concussion and that whatever injury my brain had sustained was healing on its own, I stopped worrying about my help-lessness and began concentrating on getting fit again. I began to eat and I managed to go outside to remove bark from the trunk of a nearby poplar tree so that I could scrape off the cambium pulp, which is rich in vitamin C. This I ingested raw at first, later adding stringy dollops of it to my soups. And I took to walking, at first for half an hour every morning and afternoon, then for one hour twice a day.

In this manner I recuperated to the point where I felt fit enough to go down to the Goldstream with my fishing rod to make a try for some trout. At first, casting with spoon lures, I had no luck. I was about to give up when I thought I might as well try flies as a last resort, not really expecting any strikes because it was too early in the season for insects to be found floating on or in the water. But on the third cast I hooked into a cutthroat trout that weighed about one pound; in quick succession I caught three more, then put away the tackle.

I ate two whole trout for supper, each seasoned lightly with some garlic powder and fried. For those who are accustomed to eating fish or meat on a daily basis, let me say that nothing can compare with the taste of newly caught trout after the diner has not touched a morsel of fresh protein for almost seven months! For breakfast the next morning I had the third fish, keeping the fourth for supper. After that my recovery was complete, but I waited two more days before resuming my study of the Ghost and his territory.

The evening before I was due to go up the valley in search of the tom, I found myself thinking about him almost continually, recalling the times that we had met and wondering about what he had been doing and how he had fared since I last saw him. I began to consider the country as a whole, seeing it with the puma as its focal point, but examining it as an integrated natural system that was kept in balance by every living thing that inhabited it, as well as by the geography and composition of the landscape.

Up to this point, almost my entire life had been devoted to the study of nature, a commitment I made when I was only eight years old, though I did not then realize that I had done so. My first teacher was the sea. Indeed, I was born at sea and for good measure was christened three weeks afterward by a Royal Navy chaplain on board a British warship, H.M.S. *Thunderer*, that happened to be in the harbor when the passenger vessel that was bringing my family from South Africa docked in the port of Vigo, Spain. I was raised near the Mediterranean Sea, where I played with the life forms that inhabited the shallows and watched the birds and land organisms that lived on the beaches, on and under the rocks, and in tide pools. In such a kindergarten, learning was pleasure and knowledge was an exciting thing that grew measurably on a daily basis. Before I actually entered a formal classroom, I had an elementary understanding of life and death from having repeatedly observed both of these biological facts; and I had an inherent, if as yet confused, understanding of the continuity of nature. This served me well, for when in later years and higher grades, teachers of natural science taught us to study the fundamentals of life each in isolation from the other, I persisted in searching for unity, becoming something of a nuisance because I always sought to pursue subjects beyond the limited boundaries of each lesson. After repeated tutorial rebuffs, I learned to conform in the classroom, but persisted in searching on my own, reading extracurricular textbooks, and seeking understanding in the field. With such a background, when I studied biology as an undergraduate, I developed a sort of split personality, accepting and understanding the clinical, each-part-out-of-context approach so common to most of the scientific disciplines, while the more mystical side of my personality continued to view life as an essence found in all things, which united all things and was responsible for the coordination and well-being of the environment, even to the point of ordaining that new life would emerge from every death. (This view did not, however, embrace the belief that the soul returns again to set up housekeeping in another body.)

Similarly, although the biologist in me has described the land that I was studying as being *the puma's range*, as a naturalist I recognized that the land here, as elsewhere, belonged to no individual being. The lions in the region perforce had to share it with all the other animals that lived there. Aware of these things at the start of my enterprise, I had thought it was essential to study the country and its wild inhabitants as closely as I studied the puma.

Now, as I was about to engage in the spring phase of my task, I deemed that it was time to alter my focus: to study the whole environment and to try to view in context all the varied life forms that it contained. Hitherto, I had essentially pursued a specialized subject, the puma; henceforth, while continuing to devote special attention to the cat, I proposed to become a generalist. I could not, however, expect to identify and understand every single process and activity that existed and was being continually developed within the environment, and this meant that I had to observe esthetically, rather than clinically. I would go out into the wilderness and allow it to touch me, so that I could feel its influences, share its moods, and fully appreciate its beauties outside of the restrictive precepts of science.

I could not stop thinking about the Ghost, nor did I want to, but I no longer proposed to try to analyze his every action, viewing him instead as a friend, an occasional companion whose quiet presence would always be welcome, and whose tolerance of me was gratefully accepted. In truth, I had already formed an emotional attachment to the magnificent animal, so it was easy to put aside biological considerations in favor of the enjoyment that I derived from his presence.

In such a mood, I entered the Maytime wilderness without

plans or goals, allowing interest to guide my course or to arrest my travel. I took pleasure in the profusion of new buds and shoots that decorated every plant and deciduous tree, and the songs of the birds, the new arrivals and the winterers expressing equal excitement now that the sun was warm and the nesting season had arrived.

I realize that because of my long absence from civilization, I had stepped backward in time and become a primitive, almost animalistic being. I was alone but not lonely, deriving enjoyment from everything that surrounded me and companionship from each mammal or bird that I met. My three friends, Wisa, Ked, and Jak, no longer came as a group, but as individuals, for they had by this time mated, nested, and raised broods of young ones, soot-colored offspring that resembled their parents not at all and were still too timid to come to me. But they followed one or the other of their parents through the forest, and were always quick to screech for the food I gave the adults, often making such a fuss that a male or female jay would give in and beak-feed a clamorous son or daughter that was perfectly capable of getting its own food.

Soon after I left the valley of the shack, the first mammals I noticed were three snowshoe hares that had not yet finished changing their snowy winter coats for the more subdued and practical woodsy colors of summer. Two of these were bucks and they were leaping at each other and gnashing their molars, displaying their antagonism and prowess in order to impress a large doe that squatted within the shelter of the lower branches of a young hemlock; she was quietly watching the display, but showed no interest. The bucks continued to leap, sometimes jumping six feet into the air. When they leapt at the same time, they boxed with their front legs, but did not actually hit each other. Indeed, the duel was entirely bloodless. It ended when the smaller of the two bucks quietly hopped away from the arena, whereupon the victor loped over to the doe and squatted beside her. During all this, my presence elicited no reaction from the three hares, and the remaining pair hardly bothered to look at me as I moved off. Glancing back when I was about fifty feet from them, I noticed that they were still sitting companionably, ears flat on their back, in a resting posture, nostrils quivering in time with their breathing.

During the afternoon, after having seen a variety of birds and small mammals, and while walking on the lower slopes parallel to the west bank of French Creek, I was privileged to watch as a black bear sow nursed two young cubs. The trio were below me, closer to the creek and in a fairly open location. The mother sat on the ground, but leaned casually against a tall hemlock trunk, each of her arms encircling a cub.

The little ones, born most probably in January in a mountain den, looked as though each would weigh about ten pounds. One was jet-black except for the characteristic white blaze on its chest, the other was cinnamon-brown. The breeze was in my favor and the smacking, grunting sounds made by the feeding twins had masked the slight noise of my passage, so that the mother was not aware of my presence.

I dropped onto one knee to merge my outline with that of the background vegetation and I watched through the field glasses a scene as ancient as time and as tender as any that might be enacted in a human nursery. Holding her cubs lightly with her enormous, hairy forearms, the sow looked down at each of them, licking one or the other, caressing them while they sucked greedily and as some of the milk dribbled out of their eager mouths. I would have liked to stay until the twins had finished feeding, but I didn't want to disturb the animals. After about three minutes I retreated quietly, allowing the big shaggy mother to continue to enjoy and to love her young ones unmolested.

Later, near the marsh area at the north end of the valley, I sat on a log to rest and enjoy the sunny day and the sounds of the wilderness. Overhead, a pair of bald eagles were wheeling and gliding, little more than specks against the blue sky. Near me, but unseen because they were within the trees, ravens were cawing, gurgling, and cooing, an endless conversation uttered in rather comical voices. Chickadees and nuthatches flitted from tree to tree, chatting constantly. A solitary thrush sang boisterously and melodiously from his perch on the branch of a hemlock. The afternoon was perfect, except for the presence of some early and thirsty mosquitoes that bit me when I became engrossed by some sight or sound.

I had been sitting quietly for about fifteen minutes when I began to feel that I myself was being observed. Although I resisted the temptation to look over my shoulder, believing that I was being fanciful, the sensation was so strong that I gave in to it. Turning, I was amazed to see the Ghost. He was sitting on his haunches and watching me intently from a distance of no more than forty feet, his ears cupped and held forward, his front legs braced, paws together, against the pull of the grade. The puma's long tail was curled around his paws and was absolutely still. There was no doubt that he was completely at ease; neither was there any question about his interest in me.

We looked at each other for several seconds, and then, as though to prove that his intentions were peaceful, the Ghost yawned mightily, stretching himself on the ground before he had finished the gape. Crossing his front paws, he rested his head on them the moment he closed his great mouth.

He had taken his eyes off me while he was yawning and lying down, but when I began to speak to him, he looked at me without altering his position, the attentiveness of his ears telling me that he was listening to my voice.

Without actually getting up, I turned myself right around until I was sitting facing my wild friend, and we spent some more minutes looking at each other. Then I slid off the log, stretched my legs out, and leaned against the recumbent tree, finding myself on a level with the puma and showing by my behavior that I was as relaxed as he was. A short time later he closed his eyes. Not long after that, with the low sun hot on my back, I also became drowsy.

The chill of evening awakened me and I was disappointed to find that the Ghost had gone, but this slight feeling of letdown didn't last long. Nothing could have dampened the sense of pleasure that I felt as I thought about the experience and remembered that on this occasion the big puma had voluntarily come to me, arriving as silently and ghostlike as ever, but waiting patiently for me to turn around and notice him, thereafter taking his ease in my company. Am I being anthropomorphic? Maybe, but who has not at some time or another found himself in companionable association with an animal? And who has not realized that some animals actually seek out and enjoy the company of humans? Whatever one wishes to call the contact that the Ghost and I made that afternoon, we related in friendship; and we *both* knew we were doing so.

The next time I saw the Ghost, he was stalking the marmot that lived among the rocks on the north side of my own valley, the same animal that had whistled at me in the autumn and had eluded the puma's attack.

I had been out all morning, walking along the borders of the Goldstream marsh until the newly arrived blackflies drove me away at about lunchtime, making me return to the shack, have a snack, and cover myself in repellent before venturing out again. But when halfway along my trail home I encountered the puma's tracks, I changed my mind, electing to put up with the bloodsuckers in order to follow the Ghost's spoor. He had evidently climbed up through the trees, approaching from French Creek, and had then turned on to the pathway that led to my small clearing.

At the trail mouth, I experienced a moment of *déjà vu* when I saw his big pugmarks clearly indented on the same patch of bare ground that had alerted me to his presence here during my first exploratory journey to the region. He had also paused to defecate and make a scratch mound, this one only about two feet away from the remains of the old one. Two of the tracks were in damp soil, and water was even then seeping into them, which meant that the cat had passed this way only minutes ahead of me.

By now I had encountered eighty-seven scratches left by both lions, fourteen in the female's territory and seventy-three made by the Ghost; this one brought the total to eighty-eight. In face of the evidence, and coupled with my observations here and elsewhere, I had already decided to reject the theory that holds that these marks are intended to mark a puma's boundaries against trespass by others of the species. Clearly, if the big cats heeded such "warnings," little contact would be made between them and the resulting lack of matings would have culminated in the extinction of the breed centuries earlier. I could accept that the hills might alert a puma seeking a new territory to the presence of another cat; but if the newcomer happened to be large and powerful, it would simply invade the range and dispossess the incumbent.

Similarly, since the females of the species more usually initiate contact during their time of estrus, it is logical to suppose that the scratches, far from having a "keep out" effect, are powerful lures when left by a male, and serve to inform a female in heat that she is wasting her time if the mounds are made by another female. Similarly, an amorous tom who might wander in search of a mate (which they do often enough), knows from the first sniff whether a female cat is prepared to accept his advances. If she is not, and the tom is experienced, he will know better than to dare her wrath by pressing his intentions on her.

All animals are territorial to some degree, but human observers tend to make too much of this, probably because as a species, we have the doubtful distinction of making war upon each other—something that animals don't do. Faced as we have been for many centuries by the prospect of imminent hostilities, we consider warning signs to be of major significance. In the world of animals, signs play a number of roles, the most important being the attraction of mates. Conversely, the least important is the "keep out" theory.

Pumas are loners because they must be. A pride of lions in North America would quickly kill off all the prey species and would then starve to death; but pumas are not lonely by choice. This is shown by the fact that a mother cat must actively chase away her offspring when they are ready to sustain themselves on a range of their own. The young must *learn* to live alone; *they do not come by this habit naturally.*

Thus it was my view then, and still is now, that the puma's scratchmarks are made to advertise the animal's presence to others of its kind in order to encourage sexual contact. Beyond this, the scratches may serve the territorial imperative, but no more than would ordinary stops for urination or defecation.

Looking up after examining the Ghost's tracks and the mound he had made, I saw that my shack had evidently attracted his attention. He had circled my dwelling, paused in front of the door, and then walked away, heading toward the pile of boards at the far end of the clearing. I followed, but I was still some paces from the collapsed building when I noticed slight movement within the shelter of some seedling pines. The Ghost was crouched behind the young trees, staring at something that was beyond my vision, but which was located about midway between the rotting timbers and the rocks where the marmot had its den. Here, in an open area where grasses, shrubs, and wildflowers grew, the puma's gaze rested.

I stopped as soon as I saw my friend; and I noted from the way his ears flicked backward that he was aware of my presence, but he did not otherwise react. After a moment of scrutiny I realized that something was moving among last year's dead grasses, and although I could not be positive about an identification, I guessed that it was the marmot. It seemed that last year's feud was about to be renewed. Which animal, I wondered, would lose this joust? The answer wasn't long in coming.

The Ghost, moving like a serpent, sidled forward in utter silence, flattened himself behind a boulder, moved into the shelter of a patch of sere raspberry canes, and stopped. He settled his body in the position of the charge. I still could not see the cat's target, but the quiet movement of the grasses continued, telling me that the marmot was feeding on some late-spring succulents.

The Ghost began wagging his tail more swiftly, making a whisper of sound as the appendage brushed against the grass. This immediately produced a whistle of alarm from the marmot and caused it to break cover and run as fast as its short legs could move, aiming for the burrow.

The Ghost charged, his tawny body an indistinct blur as he stretched to full length and propelled himself along by powerful thrusts of his legs, his long tail streaming behind and undulating as he ran. I didn't have an opportunity to time the action, but I am sure that only one or two seconds elapsed between the moment that the marmot emitted his piercing whistle and the instant that the puma began to move.

The marmot, looking ungainly because of its rotund body, was traveling at a fast clip; but the puma was even faster. Changing his running dash into leaps, he caught the quarry after the third jump, hitting it with a cupped forepaw. The blow was powerful enough to send the animal flying through space to land twenty feet away. The marmot lay still, dead. I judged its weight at about twelve pounds.

The Ghost turned and looked at me, and I spoke to him as usual. Then he strolled over to the kill, lowered his muzzle, and sniffed the prize before licking it. I expected him to pick up the dead animal and take it away to eat undisturbed; instead, he flopped down in front of the carcass and began chewing, not even sparing me a glance. I watched as he turned his head and pressed the right side of his face against the body of the marmot, using his carnassials, or cheek teeth, to cut through skin, muscle, and tendons.

As I stood there in the May sunshine no more than forty paces away from the Ghost, I at last found the courage to put to the test the one consuming question that had been uppermost in my mind ever since the puma had shown that he accepted and trusted me: What would he do if, as he was eating, I walked up to him and squatted before him, within his reach?

I had tried to make myself do this during a number of our previous encounters, but my nerve had always failed at the last minute. Almost as often, irritated by my lack of fortitude, I had tried to write about it in my notebook, but I always stopped before I put pencil to paper because it seemed that recording my failure would in some indefinable way aggravate my weakness and prevent me from ever knowing whether an adult, fully wild puma could be induced to accept a man so completely that he would be allowed to approach closely while the lion was feeding off a fresh kill.

The Ghost had proved often enough that he trusted me. My failure to place myself before him in a position from which there was no retreat showed that, until today, I had not fully trusted *him*. Always I had looked upon mutual trust as the most powerful bond that can exist between living things. I had on repeated occasions sought for and secured such trust by offering my own faith in return during dealings with particular animals. I was now going to do it again. I had to!

Speaking to him as I began to walk across the distance that separated us, I forced myself into a state of full tranquillity, not looking at the Ghost, but noting his presence with peripheral sight. I counted my steps. When I had taken thirty strides I glanced down, but was careful not to look directly at the lion. He had raised his head and was eyeing me. One of his great paws rested on the half-eaten carcass, the other lay relaxed on the ground, one edge just touching the marmot's head.

I could not fail to notice the blood on the tom's muzzle, cheek, and both paws; but I also noticed that his tail was not lashing and that his eyes were calm. His mouth, though partly open and showing half of the bloodstained fangs, was in repose.

I noted all these things instantly while I took six more careful strides. Now I stopped, again avoiding the Ghost's eyes; then, with infinite care, I lowered myself until I squatted only five feet in front of the big puma. I continued to speak, but more softly, as I had done many times while standing beside the bars of Tom's cage in Regent's Park Zoo; and as I did so, I heard the noise of feral teeth ripping through flesh, and the sound of chewing. And then I heard purring!

I looked at the Ghost. He continued to eat and to purr; his ears hung relaxed on either side of his broad head, but now and then his eyes lifted to my face, shining pleasantly and serene. The smell of him, rich and strong, blended in my nostrils with the smell of the kill, of the blood, and of the rank odor of the intestines and the offal; and mixing with these primal essences was the soft scent of the evergreens.

I stopped talking and listened. The Ghost kept purring and eating; I heard my own breath as it escaped my nostrils, and the voices of the birds, and the drone of mosquitoes. And I felt utter contentment and a deep affection for the great lion and for the wild environment that surrounded us.

Very quietly, I stood upright. The Ghost again looked at me with that placid gaze. I spoke quietly, but now my words held special meaning and gave my voice the sound of conviction. I turned right around and began to walk away, the Ghost's purr following me for a time.

Just before the wrecked mine building hid me from his view, I turned for one more look. The Ghost was in the act of closing his jaws on the last of the marmot's solid parts, the head.

Those few unforgettable moments during which the Ghost and I became spiritually united were to mark the end of our physical association. Soon afterward, as the caribou and goats began moving to higher ground so as to escape the flies and to seek the new succulents that day by day crept higher up the slopes, the Ghost left the valley.

I did not seek him out. That was his world, the special domain in which he was beyond the reach of civilization, free to live as his kind have done for unknown millennia.

I did not see him again. I knew that our relationship would

endure, because I shall never forget those few precious moments when the wildness of my own nature found full expression in the presence of that magnificently feral animal.

During the next week I occupied myself by dismantling the shack and taking the boards back where I had found them, for I did not want to leave a single trace of my presence in the tiny valley. Afterward, sleeping again under canvas and waiting for the Goldstream River to slow its turbulent course, I lounged a great deal, went fishing now and then, and walked a lot, confining my rambles to the lowlands of the river valley.

During these quiet wanderings I refined my knowledge of the Selkirk wilderness and met a large number of the mammals and birds that inhabit it. Some, like White Spot and his pack, were old acquaintances, while others were strangers belonging to species that I had come to know well after many years of field study. All were intriguing; each individual taught me a little more about itself while the blossoming wilderness of trees and scarps and water and icefields engraved itself forever on my memory.

On July 18, I pushed away from the French Creek sandbars and nosed the canoe into the current. This time my journey to the power line took only three hours.

Two days later, when I returned in my station wagon to pick up the canoe and equipment, I did not go down to the river, but I spent a few moments bidding a silent farewell to the Selkirk Mountains and to the Ghost Walker who lived among them.